ARRL's Wire Antenna Classics

A collection of the best articles from ARRL publications

Compiled by Chuck Hutchinson, K8CH

Edited by Dana G. Reed, KD1CW

Production Staff:
Jodi Morin, KA1JPA
Joe Shea
Paul Lappen

Cover Design: Sue Fagan

Published by:
The American Radio Relay League
225 Main Street
Newington, CT 06111-1494

FOREWORD

Wire antennas range from simple to complex, and you can probably find one to deliver the performance you want. Most hams who operate on the bands below 30 MHz have at least one wire antenna. Some have *only* wire antennas! Perhaps your first antenna was, or will be, a wire dipole.

This is a book of antennas and ideas. Dipoles, broadband dipoles, multi-band dipoles, loops, collinears, rhombics, wire beams, vertically polarized and receive antennas are all featured in this volume. There's even a chapter on trees and tree-mounted antennas.

Some of the articles are classics that were published before most of us were born. Others reflect the benefits of using the latest in computer modeling followed by physical testing.

Some articles present antennas that you may want to build and use right away. Others may give you ideas to use with your own creativity.

For RF-safety information and fundamental antenna theory, you'll want to refer to *The ARRL Antenna Book*. See the latest issue of *QST* for other ARRL antenna-related publications, or visit our online bookstore at *ARRLWeb*: **www.arrl.org/catalog/**. Please take a few minutes to give us your comments and suggestions on this book. There's a handy Feedback Form for this purpose at the back.

David Sumner, K1ZZ
Executive Vice President
Newington, Connecticut
February 1999

CONTENTS

CHAPTER ONE

DIPOLES

By James W. ("Rus") Healy, NJ2L   From *QST*, June 1991

Antenna Here is a Dipole

You've probably heard this phrase in many QSOs.
Just what is a dipole antenna, and why are they so
popular?

Dipole antennas have been widely used since the early days of radio. Simplicity and effectiveness for a wide range of communications needs are the reasons for this—and they're also the properties that make dipoles worthy of your consideration. There's more to building and installing an effective dipole antenna system than choosing the wire and insulators, as you'll see. Next month, I'll discuss choosing the right feed line for your dipole, and related subjects.

What is a Dipole?

The dipole gets its name from its two halves—one on each side of its center (see Fig 1). (In contrast, a *monopole* has a single element, usually fed against ground as a

vertical.) A dipole is a *balanced* antenna, meaning that the "poles" are symmetrical: They're equal lengths and extend in opposite directions from the feed point. In its simplest form, a dipole is an antenna made of wire and fed at its center as shown in Fig 1. (This may look familiar: You may have received a QSL card with a sketched dipole, resembling Fig 1, to denote the antenna the other station used during your contact.)

To be *resonant*, a dipole must be electrically a half wavelength long at the operating frequency. A dipole's resonance occurs at the length at which its impedance has no reactance—only resistance—at a

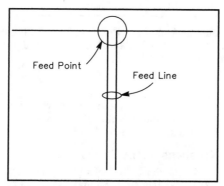

Fig 1—One of the simplest antennas used by hams, the dipole is also one of the most effective, considering the relatively small space it requires. In its simplest form, a dipole is a wire fed at its center.

Table 1

Approximate Lengths of Half-Wave Dipoles for the MF/HF Ham Bands*

Frequency	Length
28.4 MHz	16 ft, 6 in.
24.9 MHz	18 ft, 10 in.
21.1 MHz	22 ft, 2 in.
18.1 MHz	25 ft, 10 in.
14.1 MHz	33 ft, 2 in.
10.1 MHz	46 ft, 4 in.
7.1 MHz	65 ft, 11 in.
3.6 MHz	130 ft
1.8 MHz	260 ft

*General equation for half-wave dipole length: $\ell = 468 \div f$, where ℓ is length in feet and f is frequency in megahertz. This equation yields good starting points; you may have to lengthen or trim your antenna to achieve resonance. See the sidebar entitled "Dipole Construction and Adjustment."

There's more to building and installing an effective dipole antenna system than choosing the wire and insulators.

Dipoles can be installed in an infinite number of configurations other than the classical flat-top arrangement.

given frequency. As it turns out, that resonant impedance range is compatible with many common coaxial feed lines. Within limits, however, resonance isn't necessary for a dipole to be effective, as I'll explain a bit later. Resonant half-wave dipoles range in size from about 16 feet for the 10-meter band (28-29.7 MHz) to 260 feet for the 160-meter band (1.8-2 MHz). See Table 1.

The lowest frequency at which a dipole is resonant is known as its *fundamental resonance*. A dipole works best at and above its fundamental-resonant frequency. But if a total-length limitation imposed by property boundaries or the spacing of available supports keeps you from doing this, make the antenna as long as you can, even if it's not a half wavelength. Resonances repeat, for half-wavelength-long dipoles, at *odd multiples* of the fundamental-resonant frequencies. For instance, a dipole resonant at 2.5 MHz is also resonant at 7.5 MHz, 12.5 MHz, and so on. These higher frequency resonances are known as *harmonic resonances*.

As I mentioned earlier, a dipole doesn't have to be resonant to work well. More important than resonance are good construction and efficient power transfer from the transmitter to the antenna. Using an antenna tuner, you can match dipoles that are far

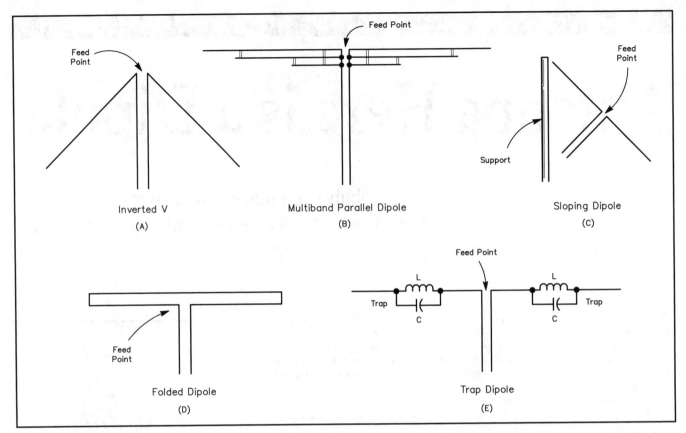

Fig 2—Variations on the dipole are numerous: at A, an inverted V; at B, a multiband parallel dipole; at C, a sloping dipole (sloper); at D, a folded dipole, and at E, a trap dipole. Dipoles of the multiband parallel, trap and folded varieties can be installed in sloping or inverted-V configurations.

longer or shorter than resonant length, but the feed line plays an important role in efficient power transfer, as I'll discuss next month. If you're primarily interested in operating on only one band, a resonant dipole is a good choice. If you're interested in multiband operation with a single antenna, the picture is a bit different. In this situation, it's a good idea to make the antenna resonant at the lowest frequency you plan to use it on (that's where the antenna is longest, because antenna length is proportional to wavelength).

Why are Dipole Antennas so Popular?

For almost any kind of MF/HF operation, dipoles are easy to build and install, and they give good results when put up at any reasonable height. "Reasonable heights" are anywhere from a few feet and up, depending on the band. A good general height guideline is half a wavelength or more, but that's impractical for many of us, especially on 40, 80 and 160 meters. At the least, a dipole should clear any surrounding buildings, and other large obstacles, for good performance.

Many hams do quite well with dipole antennas that are electrically low; for instance, an 80-meter dipole strung between two trees at 50 feet (less than a quarter wavelength) allows me to work Europeans regularly on 80-meter CW with a 100-watt transceiver. As a bonus, this antenna works

even better at higher frequencies, where it's electrically much farther above ground, as I'll also discuss later.

Variations on the Theme

Part of the beauty of dipole antennas, like many other simple things, is their flexibility. Dipoles can be installed in an infinite number of configurations other than the classical flat-top arrangement (Fig 1). Some of the more common variations include the inverted V (sometimes called the drooping dipole—Fig 2A); parallel multiband dipole

---

A dipole doesn't have to be resonant to work well. More important than resonance are good construction and efficient power transfer from the transmitter to the antenna.

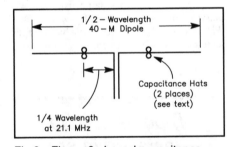

Fig 3—Figure-8-shaped capacitance hats, made and placed as described in the text, can make your 40-meter dipole resonate anywhere you like in the 15-meter band.

(Fig 2B); sloping dipole (sloper—Fig 2C); vertical dipole (Fig 1, rotated 90°); folded dipole (Fig 2D); and trap dipole (Fig 2E).

Inverted-V dipoles are probably more common than flat-top versions. As you might expect, the inverted V gets its name from its shape. The main advantages of inverted Vs are that they need only one high support, and that you can get more total wire into the same horizontal space using this configuration. This is often an important advantage on the lower- frequency bands, where real estate and sup-port height suitable for putting up a full-size dipole are at a premium. Inverted Vs usually work almost as well as horizontal flat-top dipoles when the dipole's height is the same as the feedpoint height of an inverted V. It's im-

Dipole Construction and Adjustment

Dipoles can be made of almost any kind of wire or tubing, but how well they work and how long they last is determined by the quality of the parts and care you use in building them. With that in mind, here's what it takes to make a dipole antenna.

• *Wire*. Stranded and solid copper, and copper-plated steel (Copperweld), are the most popular types used for antenna construction. Don't use soft, solid copper (such as house wire) for long unsupported runs in resonant antennas; solid wire can stretch enough after installation to detune the antenna. Use reasonably large wire (no. 16 or larger) for permanent antennas. No. 18 or smaller wire is generally okay for low-profile or temporary antennas, or those cut for 14 MHz or higher frequencies. There's little difference between antennas made from insulated and uninsulated wire although the presence of insulation generally lowers the resonant frequency of a given length of wire by a bit. Insulated wire is heavier and thicker, which, depending on its color, can make it more or less visible than its uninsulated equivalent. Davis RF carries uninsulated, stranded no. 14 copper that's made of 168 strands of wire. This is good stuff for dipoles (or home-brew balanced feed lines) because it's very flexible and easy to work with, doesn't stretch, and resists breakage well.

• *Insulators*. A dipole should have insulators (Fig A) at its center (feed point) and at each end, where ropes (or other restraints) support the antenna. The center insulator serves two primary purposes: (1) it separates and supports the feed-line conductors, making for strong mechanical wire joints that minimize stress on solder joints; and (2) it gives you a convenient place from which to support the dipole with a rope, if your dipole is center supported. End insulators serve another purpose: They provide an isolated support point at the high voltage ends of the antenna wire. (Potentials at the ends of a resonant dipole can be in the kilovolts—even with just a 100-watt transmitter!)

Plastic, ceramic or porcelain insulators are fine for most applications. Home-made varnished hardwood is also acceptable. There's no need to use a special center insulator at a dipole's feed point—the impedance, and therefore the RF voltage, there is low—although doing so can make construction easier.

• *Rope*. Nylon, Dacron and similar materials are best. Avoid polypropylene; it rapidly deteriorates in sunlight. Also avoid hemp and other natural materials. They tend to stay damp and rot in the center after getting rained on, with no outwardly visible signs of damage. If you select nylon rope, use the braided variety and carefully seal its cut ends by melting them with a match or cigarette lighter. This rope is very strong, but unlike Dacron, it also initially stretches as much as a few percent.

To make a dipole that'll stay up and that's easy to work on, using the right knots is important. Learn (or relearn) to tie a *bowline* (an unsliding loop); a *clove hitch* (for attaching rope to poles, tower legs, and such), a *tautline hitch* (a sliding knot for adjusting the tension in a rope); and a *sheet bend* (for joining the ends of two ropes). These knots are simple to tie, and are described in every edition of the *Boy Scout Handbook and Fieldbook*, as well as some merit badge booklets (look under Boy Scouts of America in the Yellow Pages for local suppliers of BSA materials). Another good reason to be acquainted with these

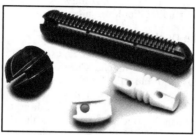

Fig A—Several varieties of insulators are commonly available to hams. The lower-center ceramic unit and the plastic insulator on the left are current Radio Shack products. The others, and close variants, are availble from many mail-order sources.

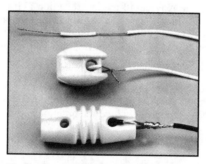

Fig B—Secure end-insulator connections are important to long antenna life. Top to bottom: Leave part of the insulation from the wire end to protect the wire where it goes through the insulator eye; wind several turns of wire back on itself to make a strong mechanical connection; solder the joint; and spray it with clear lacquer to prevent corrosion.

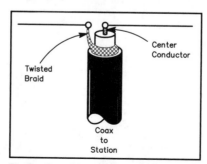

Fig C—Coaxial cable feeding a dipole antenna. The center insulator is omitted for clarity. When measuring and adjusting a dipole fed with coax, keep in mind that the stripped coax center-conductor and shield lengths *add to the antenna length*.

knots: When (if) you start working on towers (or other supports) and bigger antennas, these knots can save your life, not to mention your hardware.

• *Soldering*. To make permanent connections that will stand up to wind and weather, you'll need a soldering device that can quickly heat the wires to be joined. A 30-W iron probably won't do an acceptable job with wires larger than no. 16, a 100-W or larger soldering gun is best. If you're using relatively heavy wire and ceramic or porcelain insulators, you can use a propane torch (preferably with a soldering tip) for soldering. (If you do this, heat the joint first, then remove the flame and let the heat of the wires melt the solder into the hot junction.) Use 60/40 (or similar) rosin-core solder (available from Radio Shack and most hardware stores). Use caution when soldering close to plastic insulators; too much heat will damage them. Don't breathe solder or insulation fumes!

• *Tape, sealant and similar protective coverings*. Protect your solder joints and feed-line connections from the weather after soldering by tightly wrapping them with high-quality electrical tape (such as Scotch 33 or 88) or silicone adhesive tape, or covering them with Coax-Seal (or equivalent). Solder joints at end insulators can be sufficiently protected by spraying them with clear lacquer, available from most hardware and home-supply stores. Weatherproofing properly soldered end insulator joints is optional, but *seal the center insulator* as if you were planning to use it underwater. The last thing you want is rainwater in your coax!

In dipole construction, it's important to make strong mechanical junctions in the wires. An antenna that depends solely on solder joints to handle wind stresses will surely fall down sooner than one that's made with good mechanical connections. RF interference can also result from deteriorated solder joints. Fig B shows how to make a solid mechanical connection at an insulator.

Bringing a Dipole "On Frequency"

To resonate a dipole that's fed with coaxial cable, select a band and cut the wire a couple of feet longer than the appropriate length shown in Table 1. Then, install center and end insulators and attach the coaxial feed line (Fig C shows how to attach coaxial feed line to a dipole). Support the antenna a few feet above ground, measure the SWR and trim the antenna (equally from each end) to raise the resonance to the desired frequency. If the antenna is too short, splice in some additional wire, or attach equal amounts at the end insulators and let it hang from them. Copper-and-brass split-bolt clamps, made for attaching copper ground wires together and available from most home-improvement stores, are great for making these additions.—NJ2L

portant to keep the antenna ends high enough above the ground so that people, vehicles and such can't come into accidental contact with them.

Another common dipole configuration is the multiband parallel version. In such an antenna (Fig 2B), multiple dipole elements are fed at the same point, with a single feed line and supported by spacers attached to the longest dipole element. The main advantage of parallel dipoles is multiband coverage with resonant elements on each band, allowing the use of a single coaxial feed line for several bands without the need for an antenna tuner. An inherent disadvantage of parallel dipoles, however, is narrower bandwidth than single dipoles provide.

Dipoles are easy to build and install, and they give good results when put up at any reasonable height.

Two other fairly popular dipole variations are the trap dipole and the folded dipole. Traps are tuned circuits (consisting of inductance and capacitance) that electrically isolate the inner and outer sections of the antenna at certain frequencies, providing multiband resonant coverage from a single antenna. At a trap's resonant frequency, it presents a high impedance and therefore isolates the outer segments of the dipole, making the antenna electrically shorter than it is physically. At frequencies below the trap's resonance, it has a low impedance, which makes it transparent to RF (ie, it doesn't isolate any part of the antenna). Traps aren't used only in dipoles: Trap Yagi beams and verticals are also popular. Folded dipoles are a bit less common in Amateur Radio use; they use full-length parallel wires shorted at the ends, and have feed-point impedances that provide good matches to balanced feed lines. FM-broadcast receivers usually use folded dipoles made from TV twinlead. *The ARRL Antenna Book* and *The ARRL Handbook* cover trap and parallel dipoles in more detail.

The Dual-Band Dipole

Two popular ham bands, especially for Novice and Technician Class operators, are those at 7 and 21 MHz. As mentioned earlier, dipoles have harmonic resonances at odd multiples of their fundamental resonances. Because 21 MHz is the third harmonic of 7 MHz, 7-MHz dipoles are harmonically resonant in the popular ham band at 21 MHz. This is attractive because it allows you to install a 40-meter dipole, feed it with coax, and use it without an antenna tuner on both 40 and 15 meters.

But there's a catch: The third harmonic of the Novice 40-meter allocation (7100-7150 kHz) begins at 21,300 kHz; yet the Novice segment of 15 meters is 21,100-21,200 kHz. As a result of this and other effects, a 40-meter dipole does not provide a low SWR in the 40- and 15 meter Novice segments without a tuner.

An easy fix for this, as shown in Fig 3, is to *capacitively load* the antenna about a quarter wavelength (at 21.1 MHz) away from the feed point in both wires. Known as *capacitance hats*, the simple loading wires shown lower the antenna's resonant frequency on 15 meters without substantially affecting resonance on 40 meters.

To put this scheme to use, first measure, cut and adjust the dipole to resonance at the desired 40-meter frequency, as described in the sidebar called "Dipole Construction and Adjustment." Then, cut two 2-foot-long pieces of stiff wire (such as no. 12 or no. 14 house wire) and solder the ends of each one together to form two loops. Twist the loops in the middle to form figure-8s, and strip and solder the wires where they cross. Install these capacitance hats on the dipole by stripping the antenna wire (if necessary) and soldering the hats to the dipole about a third of the way out from the feed point (placement isn't critical) on each wire. To resonate the antenna on 15 meters, adjust the loop shapes (*not while you're transmitting!*) until the SWR is acceptable in the desired segment of the 15-meter band. You can make all these adjustments with the dipole just a few feet off the ground; raising the antenna to its permanent height shouldn't shift the SWR much. Recheck the antenna's 40-meter resonance before raising it, though.

Feed-Line Considerations

The antenna wire and insulators, how you put them together and where you string them up is only part of a dipolebased antenna system. Next month, I'll cover selecting and using a feed line for your dipole(s). In the meantime, I suggest that you have a look at the two parts of Dave (WJ1Z) Newkirk's article, "Connectors for (Almost) All Occasions," in April and May 1991 *QST* issues.

By James W. ("Rus") Healy, NJ2L

From *QST*, July 1991

Feeding Dipole Antennas

Last month, I covered dipole-antenna basics. This time I'll show what it takes to get RF from your rig to a dipole and how to make that antenna radiate as much of your signal as possible.

Communicating via radio fundamentally depends on getting RF to your antenna system—and on making the antenna radiate that energy as efficiently as possible. As simple as this may sound, achieving it can be challenging. Last month, I described how dipole antennas work and how to make them;[1] this time I'll cover getting RF back and forth between your antenna and your radio. This includes two general subjects: selecting the right feed line, and making the *antenna*—not the feed line—radiate your signals.

Feed Lines

Two general types of transmission lines are usually used to feed dipole antennas. One is coaxial cable, which is familiar to most people. Basically, it's the same stuff that connects your TVs and videocassette recorders to each other, and to the cableTV system.

Coaxial cables commonly used by hams, including RG-8 and RG-58 (both of which are somewhat different from TV coax), and similar types, are useful for feeding resonant dipoles.[2] Such cables offer a reasonably good impedance match to the antenna and to your transmitter, are easy to work with, and are fairly inexpensive. If you use coaxial cable to feed a dipole (or elsewhere in your station), follow the guidelines Dave Newkirk, WJ1Z, put forth in his article, "Connectors for (Almost) All Occasions," in April and May 1991 *QST*.

One drawback of feeding dipole antennas with coaxial cable is that most coax has relatively high loss when it's used to feed nonresonant antennas. (*The ARRL Antenna Book* and *The ARRL Handbook* discuss this in their chapters on transmission lines.) Using an antenna tuner, you can match the impedance present at the shack end of a coaxial feed line to your radio—even if it's feeding a far-from-resonant antenna—*but*

a good match isn't an indication of system effectiveness. For one thing, high cable loss causes the SWR to be lower at the radio than it is at the antenna's feed point. Therefore, if you want to use a single dipole antenna on several bands, coax isn't the best choice.

As I mentioned last month, dipoles are balanced antennas. Therefore, it's best to feed them with balanced transmission lines. In a balanced transmission line the currents flowing in the conductors are equal in magnitude and 180° out of phase, allowing the line to transfer power without radiating it. If either of these is not the case, the line will radiate, potentially causing RF interference (RFI) and other problems. Fortunately, you can avoid feed-line radiation by following the guidelines I'll cover in the remainder of this article.

Open-Wire Feed Lines

This feed-line type commonly used by hams is known variously as ladder line twinlead and parallel feeders. I'll refer to

this class as open-wire line (even though the conductors may be insulated and are separated by substances other than air). Using open-wire line to feed most dipole antennas eliminates the need for a balun (if the entire feed system is balanced, as discussed later). Open-wire lines have other advantages, especially when used for feeding nonresonant dipole antennas. When used to feed high-SWR loads such as nonresonant antennas, open-wire lines have very low loss compared to coax.[3] Even a badly mismatched open-wire line, such as one feeding a 14-MHz dipole at 21 MHz, has a lot less loss than RG-8 or RG-58 coax performing the same duty. With moderate to long feed lines, that can make a big difference when it comes to making contacts on the air. Open-wire lines are also much better for long feed-line runs than coax, because open-wire lines generally have lower matched loss—loss when operated at low SWR—than varieties of coax usually used by hams.

Characteristic impedances of open-wire feed lines are generally much higher than the 50 Ω of the coaxial cables that hams most often use. Typical ladder line, for instance, has a characteristic impedance of 400 to 450 Ω; TV twinlead, 300 Ω. To use open-wire transmission lines to feed dipoles, you'll also need to use an antenna tuner, because modern Amateur Radio gear is designed for use with unbalanced ≈50-Ω feed lines. If you're planning to use a dipole antenna on several bands, you'll need a tuner anyway, because a dipole's impedance can provide a good match to coax on only one or two HF ham bands.[4]

Widely available from *QST* advertisers, open-wire transmission lines are usually less expensive than coaxial cable and require no special connectors. If you like, you can easily make your own open-wire line with wire and home-made or store-bought insulators.[5,6]

One area where open-wire transmission

lines are less practical than coax is in routing the feeder from the antenna to the station. In properly used coaxial cable, the RF fields are contained almost entirely within the cable, so coax can be run through walls and near other conductors without special precautions. But in open-wire line, RF fields surround the line at least as far as the wires are spaced apart. Thus, when you're routing open-wire line, space it at least its width away from any conductive object, and farther if it runs parallel to a conductive object for more than a couple of feet. (Radio Shack sells standoffs meant for supporting TV twinlead adjacent to walls and roofs; these standoffs are usable with most open-wire feeders used by hams.)

There are many ways to route open-wire feeders through walls and windows. Follow these guidelines:
- Try to maintain the wire-to-wire spacing when running open-wire feeders through solid objects.
- Use good-quality, high voltage insulated wire (such as the center conductor and center insulator [dielectric] of RG-8 coax) or ceramic standoffs at walls or windows.
- Avoid following other conductors with open-wire lines.
- Seal the holes you make to keep weather and critters out.

One more caveat: TV twinlead is a generally acceptable alternative to open-wire lines intended for Amateur Radio use, but when it gets wet, it can become more lossy than coax at HF.[7]

Making the Antenna Do the Radiating

As I mentioned earlier, how well an antenna works largely depends on how much of the power put into the feed line gets to the antenna. If the feed line radiates or is significantly lossy, your antenna isn't radiating as much RF as it should. Keeping the feed system from radiating, thereby letting the dipole do its job as well as it can, isn't as challenging as you might think.[8] In fact, it's pretty easy in most cases.

If you've chosen to use coaxial cable to feed a dipole antenna, there's a good chance that the cable will radiate some of the applied signal, potentially causing interference and nasty RF burns in the shack. In most cases, a simple balun can eliminate this problem.

A balun is simply a device that interfaces an unbalanced system (coaxial feed line) to a balanced system (the dipole), providing the antenna and the feed line with the terminations they need, and keeps the feed line from becoming part of the antenna. Walt Maxwell, W2DU, put it succinctly[9] when he said that a balun's "*primary* function is to provide proper current paths between balanced and unbalanced configurations." In so doing, a balun forces RF in the feed line to flow into the *antenna*, instead of down the outside of the coaxial cable's outer conductor.

Three Simple Baluns You Can Build

Because the 50-Ω impedance of common coaxial cables (such as RG-8, RG-213 RG-58 and miniature RG-8) closely matches the impedance of a resonant dipole, the balun you use doesn't have to perform any impedance transformation. That makes the balun-building task much easier. Three effective and very simple baluns in wide use today are described as *choke baluns* because the high impedance they place on the outside of the coax keeps RF from flowing back down the outside of the cable. Each is made primarily from coaxial cable.

One drawback of feeding dipole antennas with coaxial cable is that most coax has relatively high loss when it's used to feed nonresonant antennas.

Bead Baluns

You can make an effective and simple balun using special ferrite balun cores[11] stacked over the outside of a coaxial cable.

(See Fig 1.) The cores don't affect the RF currents flowing on the cable's center conductor and the *inside* of its braid, but they stop RF current flow on the outside of the braid. Because the beads are one-piece units just large enough to slide over the outside of RG-8/RG-213 cable, put them on before you connect the cable to the antenna. Keeping this balun as close to the antenna's feed point as possible lets the balun do its job best, but it may be effective elsewhere on the line.

Once you've placed the cores over the feed line, securely tape the cores together and to the coax, so that the cores can't slide down the cable. These cores are somewhat fragile (they chip and break easily if dropped), so be careful when handling them.

A Coaxial-Choke Balun for the Shack

In his February 1990 *QST* article,[12] Rich Measures, AG6K, described the rationale for using a choke balun made from coaxial cable, and explained how to build such a balun. You can make this kind of choke balun by winding 15 or so feet of coaxial cable in a single layer on a piece of ABS plumbing pipe between 3 and 5 inches in diameter and about a foot long. (See Fig 2.) Avoid using miniature RG-8 cables (such as Belden RG-8X and Radio Shack RG-8M) and other foam-dielectric cables for this application, because the center conductor of such cable can wander through the dielectric and arc to the braid if the cable is wound too sharply. Solid dielectric cables such as RG-8, RG-213 and RG-58 aren't affected this way.

This kind of balun is pretty bulky, so it's best for use in the shack, just after the

Fig 1—A bead balun made from several ferrite beads stacked on a piece of RG-8 coaxial cable. These cores chip easily, so they should be taped together and to the cable to prevent damage. (Tape is omitted here for clarity.)

Fig 2—A coaxial-choke balun, wound from RG-58 on a piece of ABS plumbing pipe, effectively stops the flow of RF on the outside of the coax shield, but doesn't impede current flow inside the cable.

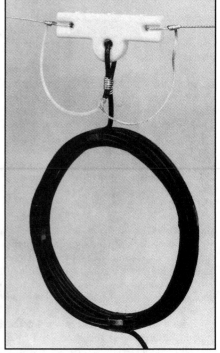

Fig 3—A coaxial-cable choke balun with an air core is effective and light enough to hang from the feed point of a well-supported dipole antenna.

output of your rig (between the radio and antenna tuner, if you use one).

A Formless Coaxial-Coil Choke Balun

Roy Lewallen, W7EL, described this variation on the coaxial-choke-balun theme in *The 1991 ARRL Handbook for Radio Amateurs* on page 16-9. This effective balun simply consists of a coil of cable, turns taped conveniently together, as shown in Fig 3. A table on page 16-9 of the *Handbook* gives the appropriate number of cable turns you should use in making such a balun. This is a function of frequency and cable type.

A coaxial-choke balun in the shack in conjunction with a bead balun or formless coaxial-cable balun at the antenna covers both bases. In fact, coaxial-choke baluns and bead baluns work comparably; use whichever best suits your situation. Also, keep in mind that all three kinds can effectively stop RF current flow on the outside of coaxial-cable braid when placed at locations other than the antenna's feed point.[13] If putting a balun at the feed point is inconvenient, try placing one elsewhere on the line. Feel free to move it to its most effective and convenient location. Choke baluns are suitable for much more than dipoles. Use them with any lowimpedance, coax-fed antennas, such as Yagis, quads, verticals and so on.

Hitching a Feed Line to Your Dipole

No matter what kind of feed line you choose for your dipole, you'll have to securely mount it to the antenna's feed point. Fig 4 shows how to attach both kinds of feed lines to a dipole center insulator. The open-wire-fed dipole insulator needs nothing more than a shot of clear spray lacquer to protect it from the elements. The coax junction, however, unprotected in Fig 4 for clarity, must be *completely sealed* for longterm reliability. Tips on these subjects are offered in the sidebar called "Dipole Construction and Adjustment" in last month's article.

Summary

No matter what you feed your dipole with, it's important to remember that the

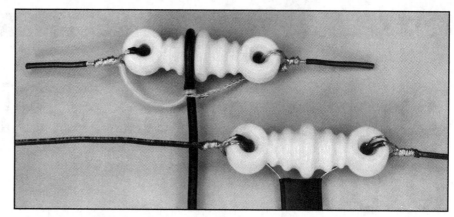

Fig 4—At left, a coaxial feed line is attached to a ceramic center insulator after being wrapped around the insulator. Securing the coax with a plastic wire tie minimizes strain on the cable. Although it's not shown here, weatherproofing is a must on exposed coaxial-cable connections like this one. At right, an open-wire trans-mission line feeds a dipole. Open-wire lines are inherently easier to use than coax because they require no weather protection except for a shot of spray lacquer at the solder joints to prevent corrosion.

dipole itself is only part of your *antenna system*. The system consists of every cable, connector and wire that comes after the radio's RF-output jack. If you take shortcuts anywhere in the system, your ability to communicate with other stations may suffer. As you'll hear many hams say, a station is only as effective as its antenna system.

Acknowledgment

Thanks to ARRL Technical Advisor Roy Lewallen, W7EL, for contributing to this article.

Notes

[1] J. Healy, "Antenna Here is a Dipole," *QST*, Jun 1991, pp 23-26.

[2] "Antenna Here is a Dipole" introduces dipole resonance.

[3] This is because, at HF, feed-line loss is mostly in the conductors (not the insulation). Because open-wire feed lines have higher impedances than coax, lower currents flow in open-wire lines at a given power level, and resistive loss is proportional to conductor current. Lower current therefore translates to lower HF loss in open-wire lines than in coax.

[4] This is true for simple single-wire dipoles, but doesn't apply to multiband, resonant dipoles such as the parallel multiband and

trap varieties described in last month's "Antenna Here is a Dipole" (see note 1).

[5] See note 1.

[6] R. Measures, "Constructing Ladder (Open-Wire) Transmission Line," D. Newkirk, Conductor, Hints and Kinks, *QST*, Feb 1990, pp 35-36.

[7] Insulators for making open-wire line are available from Davis RF, PO Box 730, Carlisle, MA 01741.

[8] See R. Lewallen, "Antenna Feed Lines for Portable Use," D. DeMaw, Conductor, Technical Correspondence, *QST*, Feb 1982, pp 51-52.

[9] In the cases of dipoles that are badly mismatched to the feed line (such as very short antennas), especially lossy antennas, and other unusual circumstances, making a dipole radiate most of the applied RF can be dfflicult. Garden-variety dipoles, however, tend to be very efficient—usually well over 95%.

[10] W. Maxwell, "Some Aspects of the Balun Problem," *QST*, Mar 1983, pp 38-40.

[11] These cores are available from Amidon Associates as part numbers FB-77-1024 (for RG-8-size cables) and FB-73-2401 (for RG-58-size cables). See *The ARRL Handbook for Radio Amateurs* for more information on ferrite-bead selection for this application.

[12] R. L. Measures, "A Balanced Balanced Antenna Tuner," *QST*, Feb 1990, pp 28-32.

[13] Roy Lewallen private correspondence, May 8, 1991.

The Inverted V-Shaped Dipole

With sunspot activity on the skids, the 40 and 80-meter bands are going to assume increasing importance in DX work over the next few years. The simple antenna described here has been giving a good account of itself in many installations for both long- and short-haul work.

F or the past eight years, the author (and several others at his suggestion) have been using a type of antenna that has consistently brought better signal reports on 80 and 40 meters, in comparative tests, than more conventional types such as the ground-plane and horizontal and vertical dipoles. Furthermore, it actually costs less and is easier to put up than most other types commonly used for these lower-frequency bands. Other advantages are that it can be put up in a smaller lot than required for a horizontal dipole, and the antenna does not have to support the weight of a feed line, which is quite a consideration where co-axial line is used.

Resonant Length

Fig. 1 shows the simplicity of the inverted V-shaped dipole. It consists of a half-wave dipole supported at the center, with the two halves dropped downward at an angle from the horizontal. Sloping the wires in this manner causes an increase in the resonant frequency so that a somewhat longer length of wire (approximately 5%) is required for the same frequency. However, the resonant length will be influenced by other factors in each individual case, so the length should be adjusted experimentally for each installation. This can be done with an s.w.r. bridge in the feed line, the length of the antenna being adjusted for minimum s.w.r. at the desired frequency.

Impedance and Band Width

Sloping of the wires also results in a decrease in the feed-point impedance. A 50-ohm line will usually give a closer match than a 70-ohm line. While the angle of slope is not critical, it will be found that as the angle between wires becomes sharper, the

Q increases and the bandwidth is narrowed. This narrowing can be limited by using three- or five-wire conductors or "cages" rather than single wires for the antenna (see photographs).

Directional properties are not pronounced, although there is some slight emphasis at right angles to the direction of the wire.

Two-Band Operation

For 40-meter operation, a separate similar dipole may be used. It may be connected in parallel with the 80-meter dipole at the

feed point and both may be fed with a single coaxial line. The 40-meter dipole may be run in any direction relative to the 80-meter dipole, but if the two dipoles are run at approximately right angles, as shown in Fig. 2, they will have less interaction and may also be used as the upper set of guy wires for a mast support.

Support

As with any other type of antenna, the inverted V-shaped dipole should be elevated as high as possible. It is quite feasible to

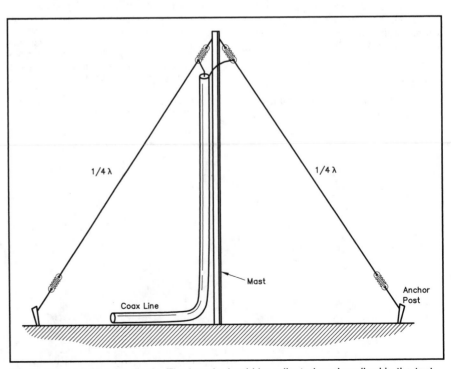

Fig 1—The inverted-vee dipole. The length should be adjusted as described in the text.

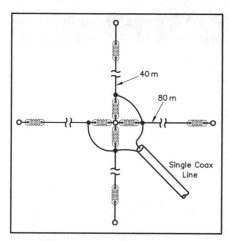

Fig 2—Top view of a two-band arrangement. Dipoles for 80 and 40 meters are connected in parallel and fed with a single coaxial line.

Details of cage construction for broadbanding elements. Spreaders are paraffined wood. [An exterior-grade polyurethane finish could be used to seal wood, or you could substitute lengths of PVC pipe for the wood.—*Ed.*]

K7GCO's antenna is shown here supported by a tall tree. Multiwire elements are used to increase bandwidth.

> *It can be put up in a smaller lot than required for a horizontal dipole, and the antenna does not have to support the weight of a feed line.*

use a tree as a support since most of the branches will be near the low-potential portion of the antenna. The elasticity of nylon cord makes it a desirable material for anchoring the ends of the dipoles. And, if a tree is used, the time-tested system of pulleys and counterweights may be used to advantage.

A tower or pole supporting a beam antenna for the higher frequencies has been used as the center support for this antenna with no apparent impairment of the performance of the beam. However, it is probably a good idea to keep the apex 5 or 6 ft. below the array.

Feeding

In feeding this antenna, the same transmission-line problems must be considered as with any other antenna. Although coax feed can and has been used, the workable bandwidth of any system using coax feed is limited if losses from a high s.w.r. and problems in loading are to be avoided. The author prefers tuned open-wire line not only because losses when working over the full width of the band are minimized, but also because it maintains a balanced system.

Results

In numerous tests in which it was possible to switch antennas instantly, the inverted V-shaped dipole has invariably proved to be superior to a half-wave horizontal dipole at the same height, a vertical dipole and a ground plane, which were used for comparison. It is assumed that the sloping results in a lower angle of radiation. K7GCO, running 600 watts input, has been frequently reported by DX stations as one of the top signals from the W7 area on the 40-meter phone band. DX on 75 meters includes an S8 contact with EL4A in Liberia–a fair haul from Washington on any band. Others who have tried this antenna have reported similar results, while some have found it the answer in covering shorter distances (100 to 1000 miles) where both vertical and horizontal antennas had previously been required to assure reliable coverage.

Some work has been done at K7GCO on a fixed-direction beam using two elements of the inverted-V type, one as a director, supported at the ends of a 15-ft boom on a tower. Results so far have been encouraging on both 40 and 80, although the spacing is rather close for 80.

An Easy-Up and Easy-to-Store Field Day Dipole

For many years, the Mid-Missouri ARC (Jefferson City, Missouri) has used several 80/40-meter dual-band dipoles for its Field Day stations. While the antennas worked well and were relatively simple to install each Field Day, we found it difficult to pack and store them. The separate wire elements (four of them) often became badly tangled and it was difficult to adequately handle the center insulator as the antenna was rolled up. We tried several different devices for rolling up the antennas, but none were completely satisfactory.

This past year, we decided to construct a set of new dipoles. This time they're made from ladder line, and we've found a better way to store them.

Constructing the Antenna

In an effort to conserve club funds, I found that a full-sized (132-foot) dual-band 80/40-meter dipole could be cut from a single 98-foot length of ladder line. I chose high-power #16 multistrand Copperweld ladder line for the project because it provides acceptable wire separation, strength, durability and flexibility all in one package.

The legs of the dual-band dipole are created by measuring 33 feet in from *each* end of the 98-foot length and cutting one wire at each end. (See Figure 1.) Then remove all plastic spacers *between the two cuts*. The result is two equal lengths of ladder line, each containing one 65-foot leg and one 33-foot leg. Once the two sets of dipole elements are cut to length, the two wires that will be at the dipole center are stripped for about one inch, twisted together and soldered.

Figure 2 shows the construction of the center insulator. There are three ¼-inch holes in a 4×3-inch piece of ¼-inch-thick Plexiglas (scrounged from the "junk box" at our local hardware store). Each hole is ⅝ inch from its respective side. The three

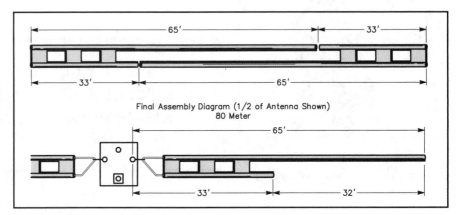

Fig 1—Cutting diagram for making an 80/40-meter dipole from 98 feet of transmitting ladder line.

Final Assembly Diagram (1/2 of Antenna Shown)
80 Meter

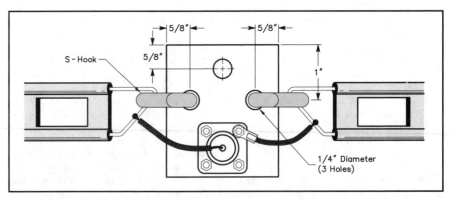

Fig 2—Details of the center insulator.

holes accept the dipole elements (on the sides) and a support rope (at the top). Drill a ⅝-inch diameter hole near the bottom to accept an SO-239 socket.

Instead of merely attaching the dipole wires through the holes, as one would expect, I installed 1½-inch S-hooks in each of the side holes. Once in place, I closed

the S-hook ends containing the plastic.

Short lengths (3½ inches) of #10 (fine) stranded wire connect the two terminals of the SO-239 to the dipole legs.[1] The wires are cut a bit long for added flexibility. I stripped about ½ inch of insulation from the free end of each wire and formed the ends into small U-shapes. They loop *loosely* over the center ends of each of the dipole elements, each is placed off to one side, clear of the S-hook and soldered in place. I was careful to form the wires loosely around the antenna wires because I wanted them to be easily desoldered and removed for field repairs.

I wanted to fold the dipole elements together at the center insulator and wind them on the roll as one unit. This requires that the center ends of the dipole elements be very flexible, so they wouldn't break after a few storage cycles. That's why I installed the S-hooks in the center insulator. The S-hooks permit easy assembly and disassembly of the leg-to-center-insulator connection for installation and storage. The leads from the coax connector to the dipole elements are long enough that the center insulator can be completely detached from the dipole center ends but remain connected to the socket. This permits the center insulator to be located outside of the reel, while the ladder line is contained inside the reel.

Constructing the Reels

I wanted to find (or build) a reel on which to roll the antenna. Jeff at Davis RF,[2] offers surplus wire reels at a reasonable price, but they're too wide for individual dipoles. I planned to disassemble the reels and rebuild them so they're barely wider than the 1-inch ladder line.

I obtained several surplus reels from Davis RF and engaged my friend Joe Betros, KBØQHZ, to do the "hard" work. Then I designed, and Joe built, a wonderful reel for the new dipoles. (See Figure 3.) The reel features are:

• Lightweight, less than two pounds

• Small size, 11-inch diameter, only two inches thick

• Durable, plywood sides and heavy cardboard center

• Hardware friendly, the four-inch-diameter center spool prevents sharp bends in the wires. The setup easily accommodates the dipole center insulator

Joe reworked the wire reel into an antenna reel: He first disassembled the reel (it was bolted together). He then cut the cardboard center spool from 12¼ inches wide to 1½ inches (1-inch ladder-line width + ¼ inch of slop + allowance for the channels in the reel disks that hold the center spool).

Joe cut a ½-inch-wide radial slot in one of the reel disks and a one-inch-wide slot all the way *across* the center spool (that is, along the tube's length). The two slots are aligned when the reel is reassembled. In use, the center insulator hangs outside the reel center, the ladder line is fed through the slots so that it's between the disks and the line is then rolled onto the reel.

Joe then applied three coats of varnish to all exposed reel surfaces. We were worried about the durability of the cardboard tube in wet weather, but the varnish removed all doubt. We are pleased with how

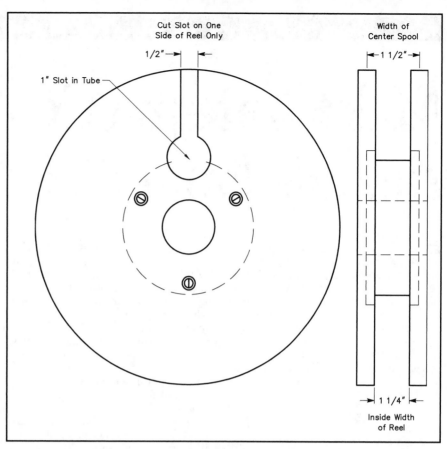

Fig 3—Views of the wire reel after modification for storage of a ladder-line dipole.

sturdy and well coated it is. Once the varnish was dry, Joe reassembled the reel. The finished product should last us for *years!*

Tune Up

With the dipole and reel complete, we slipped the center ends of the dipole over each open end of its S-hook, tossed a rope over the nearest tree and proceeded to cut the dipole elements to frequency. As cut, you should find that both dipoles are too long (too low in frequency) and will require trimming to resonance. Nevertheless, it's easier to cut them shorter than make them longer.

We use the venerable "one-cut" method of dipole tuning:

1. Raise the antenna and measure its resonant frequency (say 3.500 MHz).

2. Use the resonant frequency (in megahertz) in the standard formula to determine the antenna's effective length (468 / 3.5 = 133.7 feet).

3. Use the formula *again*, with the desired resonant frequency (say 3.650 MHz; 468 / 3.650 = 128.2 feet).

4. Subtract the length found in Step 2 from that found in Step 3 to obtain the needed total length change (128.2 – 133.7 = –5.5 feet). Add (positive result) or subtract (negative result) one-half of the length at each end of the antenna.

Of course, in actual use, you can cut your dipole elements for any desired band. Although you can calculate lengths for any frequency, I find it best to cut the dipole legs a bit longer than the calculated length and then do the final pruning on the air.

(I recommend mounting the end insulators a few feet inside the calculated length and letting the excess wire hang. That way you can easily prune the hanging ends to resonance without moving the insulators.— *Ed.*)

There are several benefits from making our dipoles out of ladder line: We needn't spend Field Day set-up time untangling four single-wire dipole elements. The elements now unroll easily, relatively straight and untangled. (Ladder line sometimes has some natural twist that appears when it's in the air, but we find this no problem.) We can now pack our dipoles in about 25% of the time it once took. We no longer need to support four separate dipole ends. With ladder line, we only need two supports.—*Tom Hammond, NØSS*

[1]While it's common to operate portable and emergency antennas without baluns, consider adding a balun to any such antenna.

[2]Davis RF Co, PO Box 730, Carlisle, MA 01741; tel 978-371-1356, 978-369-1738, 800-328-4773 (Orders only), fax 978-369-3484; e-mail **davisrfinc@aol.com**; URL **http://www.cqinter-net.com/davisrf/.**

Make Antenna Center Insulators From PVC Pipe Caps

Fig 1 shows three versions of a center insulator for a coax fed dipole. The center insulators are made from PVC slip caps, with the coax connectors and eyebolts mounted directly into the cap walls. These insulators are small, strong, lightweight and easy to build. They are built in 1¹/₄-inch caps (outside diameter is about 2.0 inches). Close the open faces of the caps by gluing on a disc of sheet plastic. This sheet doesn't carry any load, it just seals and waterproofs the center insulator.

You can use several kinds of plastic sheet here. PVC sheet (gray in the photo) is available from hobby stores. Glue it to the PVC cap with PVC cement or vinyl glue. White polystyrene sheets are also available from hobby stores; glue them with coil dope from electronics distributors or with universal plastic cement (methylene chloride) from hobby stores. This cement will also attach Plexiglas sheet—available from hardware stores—to the PVC cap. This was done with the smaller center insulator in the photo, made from a 1¹/₂ inch cap.

Use stainless steel or brass eyebolts and hardware for these insulators; plated steel will corrode. You can use PVC caps as manufactured, but the ones in the photos were cut down to an inside depth of about 1¹/₁₆ inch, which comfortably holds the coax chassis-mount connector. Notice that an area on the side was filed flat to mount the coax connector.

Coax connects at the bottom of the in-sulators, and the antenna wires fasten to the two eyebolts at the sides. Stranded wire connects from the coax socket to the antenna wires. In two of the examples, the stranded wires exit the cap through holes beside the eyebolts. I sealed the wire insulation to the cap where it exits the holes. In the third example, the wires connect the coax socket to solder lugs on the eyebolts *inside the cap*. Additional lugs and wires on the eyebolts *outside* the cap connect to the antenna wires. In all cases, wrap the stranded wires securely around the antenna wires, solder them and protect the joint against the weather. The eyebolt at the top is for an optional center support, as for an inverted V.—*Robert H. Johns, W3JIP*

W3JIP's idea suggests other applications for this technique. For multiwire dipoles, you could use a larger pipe cap (say, 4 inches, with pairs of eyebolts around the circumference for each band) or a section of pipe. See Fig 2 for some ideas.—*KU7G*

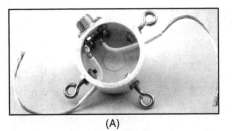

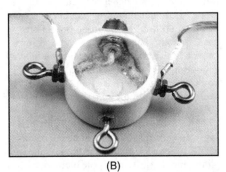

(C)

Fig 1—Three examples of center insulators. At A, the cover is left off so that the interior is visible. At B, clear polystyrene covers the opening and the interior is somewhat visible (Dust from sanding the cap clings to the cover and interior parts, obscuring the view.) C is a completed insulator with a gray PVC cover sheet glued in place. See text.

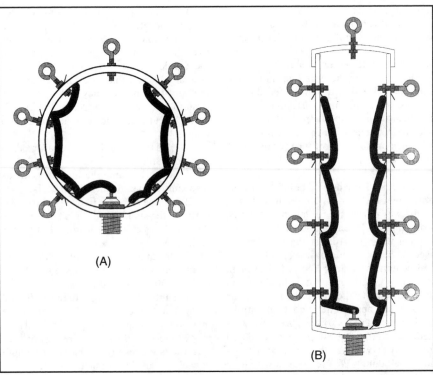

(A)

(B)

Fig 2—W3JIP's idea applied to multiwire dipoles. A is made from a 4 inch PVC pipe cap. B is a cross section through a section of 2 inch PVC pipe.

By Bill Wright, GØFAH   From *QST*, February 1996

Four Bands, Off Center

Every dipole must be center fed, right? Not necessarily. Here's a dipole that's fed *off center*, works on four bands, and doesn't require an antenna tuner.

Have you ever wondered why you're required to attach your feed line to the center of a dipole antenna? The middle is a good place for a half-wavelength antenna because the feed impedance is low, typically close to 50 Ω, when the antenna is cut to resonance at the operating frequency. This makes for a good match for 50-Ω coaxial cable, and a good match for your radio. But could you get a more versatile antenna by moving the feedpoint *away* from the middle?

Like many amateurs, I often ask other hams to describe the antennas they're using. Most of the time the answer is a wire antenna like the classic half-wavelength dipole. On occasion, however, some European amateurs tell me that they're operating with "FD3" antennas. Being more than a little unfamiliar with this design, I was eager to find out more.

After some research I learned that the FD3 is a single-wire antenna, with the feedpoint not in the middle, but *one third the way from one end*. It's coax fed with a 6:1 balun at the feedpoint. It actually resembles the Windom antenna with the single-wire feed that was popular in the early 1930s.

Studying the FD3 gave me an idea for the antenna shown in Fig 1. This off-center-fed dipole works on four bands: 40, 20, 15 and 10 meters. And, as a bonus, you don't need an antenna tuner! This antenna is similar to the 1950s Windom antenna that was fed with 300-Ω twinlead.

Construction

Imagine that you have 69 feet of #12 copper wire. If you were to cut this wire in two equal halves and feed it with 50-Ω coax, you'd probably find that it is resonant at the bottom end of the 40-meter band. (This depends, of course, on how high the antenna is above ground and so on.)

For your nonimaginary antenna, use 69 feet of bare copper wire, but *don't* cut it into equal halves. Instead, cut one length at 23 feet and the other at 46 feet. Rejoin

the two sections with an insulator in between. This off-center feedpoint will have an impedance close to 300 Ω when you apply a 40-meter signal. This same feedpoint will also present a 300-Ω impedance on the 20 and 10-meter bands, at a typical height of 40 feet or more.

Connect ladder line, either the 300 or 450-Ω variety,[1] at the feedpoint. At our one-third feedpoint, the impedance will be very high on the 15-meter band. But if you make the ladder line a quarter wavelength long at 15 meters, it will transform the high impedance at the feedpoint down to a low impedance near your radio.

A quarter wavelength of 300-Ω ladder line is about 10 feet for 21 MHz. This is probably going to be a little short to reach your radio. You can make it longer on one condition: The overall length must be an *odd* multiple of the 21-MHz $^1/_4$ wavelength. For best SWR on all four bands, I recommend

[1]300-Ω ladder line is preferred for this design, but it isn't commonplace on your side of the Atlantic. Call around to several *QST* advertisers who specialize in wire and feed lines and you'll probably find it. If not, don't hesitate to use 450-Ω line.

mend either 55 or 111 feet of 450-Ω line (or 50 or 110 feet of 300-Ω line). One of these lengths should get the ladder line to your radio with room to spare.

Now that we have a low impedance at the end of the feeder, we use a 4:1 or 1:1 balun to make the transition to 50-Ω coax. Use a 4:1 balun for 40, 20 and 10 meters, and a 1:1 balun for 15 meters. At my station I have 4:1 and 1:1 baluns that I can plug in as required when I change to and from the 15-meter band. You can purchase 1:1 and 4:1 baluns from a number of *QST* advertisers.

Conclusion

How well does the off-center dipole work? I enjoy the convenience of hopping from one band to another without having to fiddle with a tuner. I found that changing the balun when moving to the 15-meter band wasn't all that cumbersome. By choosing the correct line length, the balun was right next to my radio. Best of all, the SWR never exceeded 2:1 on any of the four bands.

Not only is the antenna easy to use, it rewards me with plenty of contacts. Off center, yes, but *spot-on* performance!

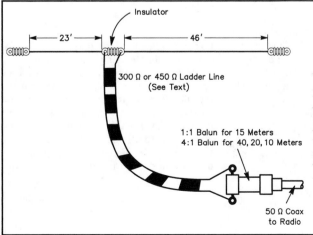

Fig 1—This off-center fed dipole offers four-band performance without an antenna tuner. Just cut the wires and the ladder line to the proper lengths. You'll need to swap baluns when you want to operate on 15 meters.

Insulator · 23' · 46' · 300 Ω or 450 Ω Ladder Line (See Text) · 1:1 Balun for 15 Meters · 4:1 Balun for 40, 20, 10 Meters · 50 Ω Coax to Radio

An 80-Meter Receiving Antenna Using a Rotary Dipole

◊ If you are using a vertical or other noisy antenna on 80 meters, you may benefit from using a separate receiving antenna. Because most of us lack the space for a Beverage, we often try various schemes for improving the S/N ratio, wishing we had room for a high, horizontal dipole! If you have a rotatable horizontal dipole for use on 30 or 40 meters (I own a Cushcraft D3W), you can tune it for use as an 80-meter *receiving* antenna. It will provide the same S/N ratio as a full horizontal dipole at that height, but with less gain.

Using a pair of SPDT relays (see Figure 1), I switch the feedpoint of my Cushcraft D3W dipole through a pair of series tapped coils, each having an inductance of approximately 20 μH. This allows me to resonate the dipole at 3525 kHz. Because the dipole's feedpoint impedance appears to be about 12 Ω, I employ a 4:1 matching transformer to step that impedance up to nearly 50 Ω. A 4:1 current balun using two toroids, described by W7EL,[1] works well. A toroid with 1:2 turns ratio should also work nicely. (Coil adjustment can be done using a 30 pF capacitor as a substitute for the D3W antenna.)

I placed the relays close to the dipole feedpoint to minimize detuning it on its native bands. Even so, I found it necessary to compensate by shortening the inner sections of the dipole about 3 inches on each side. I use a PVC box (available at electrical supply stores) to house the relays, tapped coils, the balun normally used with the dipole, and a matching transformer. The

[1]Gerald Hall, K1TD, Paul Rinaldo, W4RI, Maureen Thompson, KA1DYZ, *The ARRL Antenna Compendium Vol 1*, (Newington: ARRL), 1985, pp 163-164, Fig A3-4.

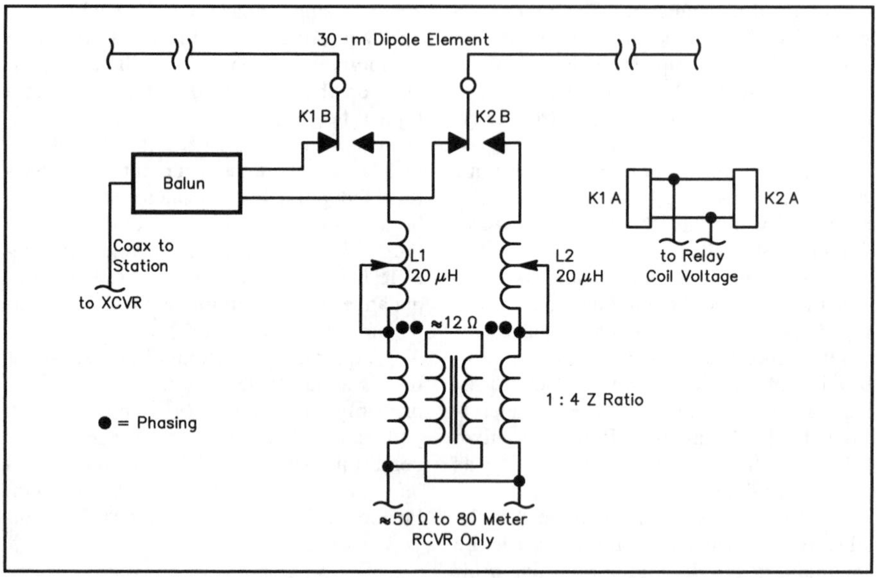

Figure 1—W6YA's approach to using a 30-meter rotary dipole as an 80-meter receiving antenna. The relays are mounted in a weatherproof box close to the dipole feedpoint. After readjusting the dipole length for 30 meters, tune the matching network on 80 meters by adjusting the L1 and L2 coil taps with K1 and K2 in the 80-meter position (energized).

K1, K2—SPDT relays with high-current-handling contacts; appliance relays used here.

L1, L2—20 turns No. 18 wire air-wound; 2 in. dia, 1¼ in. long (approximately 20 μH).

L3—1:4 impedance ratio matching transformer (see referent of Note 1). Type 77 ferrite toroidal core with 1:2 turns ratio.

box is mounted on the mast immediately below the dipole using a single **U** bolt.

After using the receiving dipole for one season on 80-meter CW, I have stopped using the 80-meter transmitting antenna for receiving except after a rainy period when the noise level is low. Signal loss is typically about 15 dB and would be less with a larger 40-meter dipole. I do use a receive preamplifier. After installation, I was able to null a local power line noise source 25 dB by rotating the dipole. Even without the null, there is a significant S/N improvement of two to five S units.

If you already have a horizontal antenna on 80 meters, don't waste your time with this—you are luckier than most. If you are using a vertical, inverted **V** or other noisy antenna, give this idea a try. I think you'll find it worthwhile!—*Jim McCook, W6YA*

By Allen B. Harbach, WA4DRU, VP5AH, VP1AH   From *QST*, December 1980

Broadband 80-Meter Antenna

The cage is back! Almost forgotten since the 1920s, this multiwire antenna, arranged as a center-fed dipole, provides edge-to-edge band coverage without the help of a tuner. The low SWR will make you and your rig happy!

I dislike antenna tuners! I suppose there is a place for them when one can put up only one piece of wire to cover all bands, but they definitely slow down the ability to QSY quickly from one end of the band to the other to catch the rare one.

When I began chasing DX in the early '70s, I rapidly became aware that something had to be done to broaden the response of my antenna system — particularly on the 80-meter band. The reason 80 meters is so tough is that it has the greatest percentage bandwidth of any of the popular amateur bands (see Table 1). Percentage bandwidth is a concept that gives a clue to the required Q of an antenna in order to have low SWR from top to bottom. It is calculated by dividing the bandwidth (in kHz) by the band-center frequency (in kHz) and multiplying by 100 to get percent. The 80-meter band is 13.3% wide

$$\left(\frac{500}{3750} \times 100\right)$$

which means that it requires an antenna Q of 7 or below to be able to cover the whole band at low SWR. To further illustrate the concept, the 15-meter band is nearly as wide as the 80-meter band, in kHz, but is much narrower in percentage bandwidth

$$\left(\frac{450}{21,225} \times 100\right)$$

An antenna Q of 45 or less will cover the entire 15-meter band with reasonable SWR (a dipole of no. 12 wire). On 80 meters the typical dipole of no. 12 wire has a bandwidth of 75 kHz at the 1.5:1 SWR points,

or in excess of 5:1 at the band edges when resonated at 3750 kHz (band center).

The higher the antenna, the better it is for DX.

To get around this problem, I did some reading in the library at the local engineering college. I arrived at the well-known fact that the fatter one makes an antenna, the lower the Q; hence the greater the bandwidth. But how thick? Doing some more

reading and a lot of paper scratching, I arrived at some relationships that could be solved with the average scientific calculator. Later, I programmed these into our company computer to speed calculations and print tables and graphs.

The equations and math I'll tackle later for those who are interested. For the others who want to know what to build, I'll cover that now. Calculating several antennas from the equations, I found that the antenna had to be at least 3 feet in diameter to cover the whole 80-meter band with a low SWR. Now, how to put up a 3-foot-diameter pipe 120 feet long! That's the question. So back to the books!

More reading showed that one can approximate a cylindrical conductor with parallel wires of various configurations. The equivalent diameter of a conductor, made up of parallel wires, is shown in Fig. 1.

Table 1

Percentage Bandwidths for the Popular Amateur Bands

Band (meters)	160	80	40	20	15	10	6	2
Percent Bandwidth	10.5%	13.3%	4.2%	2.5%	2.1%	6.3%	7.7%	2.7%

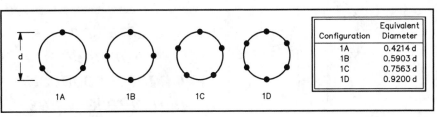

Configuration	Equivalent Diameter
1A	0.4214 d
1B	0.5903 d
1C	0.7563 d
1D	0.9200 d

Fig 1—Solid-tube equivalents of wire cages.

The reason 80 meters is so tough is that it has the greatest percentage bandwidth of any of the popular amateur bands.

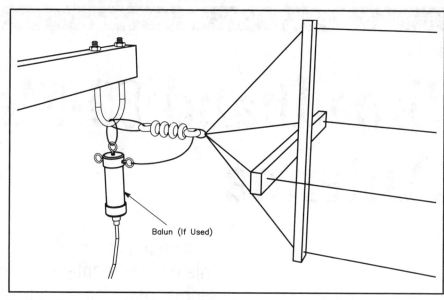

Fig 3–Center-support and end-taper detail of the cage.

Table 2
Characteristics for the 80 Meter Band

Freq.	Ohms	Reactance	SWR
3.500	53.4	− 45.0	2.18
3.520	54.2	− 41.3	2.03
3.540	55.0	− 37.5	1.90
3.560	55.8	− 33.8	1.78
3.580	56.6	− 30.1	1.67
3.600	57.4	− 26.4	1.56
3.620	58.2	− 22.7	1.46
3.640	59.0	− 19.0	1.37
3.660	59.8	− 15.4	1.29
3.680	60.6	− 11.7	1.21
3.700	61.4	− 8.0	1.14
3.720	62.2	− 4.4	1.07
3.740	63.0	− 0.7	1.02
3.760	63.8	2.9	1.06
3.780	64.7	6.6	1.12
3.800	65.5	10.2	1.18
3.820	66.3	13.9	1.25
3.840	67.1	17.5	1.33
3.860	67.9	21.1	1.40
3.880	68.7	24.8	1.48
3.900	69.5	28.4	1.56
3.920	70.3	32.0	1.64
3.940	71.1	35.7	1.73
3.960	71.9	39.3	1.82
3.980	72.7	42.9	1.91
4.000	73.5	46.5	2.01

Note: Calculations for an antenna 124 feet (37.8 m) long and 3 feet (0.9 m) in diameter covering 3.5 to 4.0 MHz. Z_o = 62.

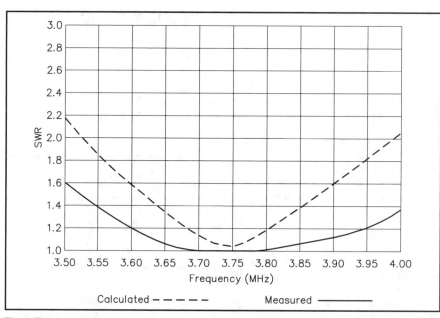

Fig 4–This graph shows the calculated vs. measured SWR values for the broad-band cage antenna over the entire 80-meter band. The gradual slope of the measured curve and the low SWR range indicate good bandwidth and matching.

The easiest type to construct is a four-wire cage. For my antenna, I used cross sticks of 1 × 1 material, 4 feet (1.2 m) long, which were held together at the center by a couple of brads. Holes were drilled in the ends of the sticks to take the antenna wire. Wire ties served to keep the spreaders from slipping (Fig. 2). I used a no. 16 wire for each element. This is equivalent in antenna resistance to a dipole made of no. 10 wire, and it keeps ohmic losses low.

Mechanical Considerations

Some mechanical considerations must be kept in mind. This antenna will swing in the wind. The first antenna I installed failed through fatigue both at the center and the end points. Therefore, the end sections of each half must be made of heavier material. I have used both no. 16 Copperweld and no. 12 soft-drawn copper wire for the end sections with no failures in over six years.

The ends of each half section are tapered over a distance equal to the spreader length to provide a transition between the large-diameter conductor of the antenna and the balun

One can approximate a cylindrical conductor with parallel wires

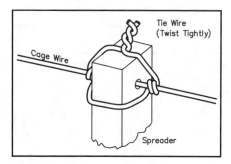

Fig 2–Detail of spreader ties.

or coaxial connection. To keep construction simple, I did not attempt to optimize the end terminations.

Use fairly heavy insulators at the center support, as this is a heavy antenna; wind loading is five times that of the usual dipole. (Do not despair! Mine has survived a twister and a hurricane!) I used a separate insulator for each half with each fastened to a U bolt in a wooden arm protruding from my tower (Fig. 3). Separate insulators at the center allow each half to be made and raised separately. A no. 12 flexible wire connects the center of each half to the balun or coaxial line.

Naturally, the higher the antenna, the better it is for DX. Mine is 68 feet (20.7 m) at the center, with one end held at 55 feet (16.8 m) and the other at 40 feet (12 m) above ground.

Testing

Once in place, the antenna is ready for testing. Each installation seems to have its own peculiarities, the result of nearby objects such as trees, houses and metallic structures. These affect the resonant length of the antenna to a greater extent than they would affect a single-wire dipole because of the larger capacitance between the antenna and nearby objects. While the length was calculated to be near 124 feet (37.8 m), I had to shorten mine to 115 feet (35 m) to have it be resonant at the center of the band. I performed the shortening in the last outboard section rather than redo the end termination. That, however, was a personal choice.

All the theory in the world is useless if the thing doesn't work! I'm delighted to say, though, that the antenna does perform well. Observe, for instance, the calculated SWR plot and the measured SWR curve in Fig. 4. The return on invested time is very high. It took me only one afternoon to put

the thing up. My rewards for the 80-meter portion of both 5BDXCC and 5BWAS were gained with the use of this cage dipole. The significant advantage of this antenna, however, is that you can throw away that 80-meter antenna tuner and QSY all over the band with ease without concern about the SWR!

Math 'n Stuff

The characteristic impedance of an antenna with a length-to-diameter ratio greater than 15 is given by the expression

$$Z_{in} = R(k\ell) - j[120 (Ln\ 2\ell / a - 1) \cot(k\ell) - X(k\ell)]$$

where
2ℓ = total length
a = conductor radius
$k\ell = 2\pi (\ell/\lambda)$, or the length of one half the antenna measured in radians.
Ln = natural logarithm

Since $\lambda = 984.25/f_{MHz}$, then $k\ell = 6.384 \times 10^{-3} f_{MHz}\ell$, where ℓ and λ are in feet.

$R(k\ell)$ and $X(k\ell)$ are quite complex functions, but are calculated as a table in Ref. 1. Fortunately, we are interested in antennas near $1/2$ wavelength long. In this region, these functions can be approximated by the following linear equations:

$R(k\ell) = 102 (k\ell) - 87.86$
$X(k\ell) = 48.54 (k\ell) - 34.86$

Some error is introduced by this approximation, but it is less than 5%. Antenna location, height and trees will introduce larger errors than that! Now, the equation for the center impedance is simplified to the point where one can calculate values with the average scientific hand-held calculator.

For angles calculated in radians:

$Z_{in} = (0.6512\ f_{MHz}\ \ell - 87.86) - j\ [120$
$(Ln\ 2\ell/a-1) \cot (6.384 \times 10^{-3}\ f_{MHz}\ell)$
$- 0.3099\ f_{MHz}\ell + 34.96]$

For angles calculated in degrees:

$Z_{in} = (0.6512\ f_{MHz}\ \ell - 87.86) - j\ [120$
$(Ln\ 2\ell/a-1) \cot (0.3658\ f_{MHz}\ell)$
$- 0.3099\ f_{MHz}\ \ell + 34.96]$

SWR calculated by the *Antenna Book* formula:

$$SWR = \frac{1+k}{1-k}; k = \frac{(R - Z_o)^2 + X^2}{(R + Z_o)^2 + X^2}$$

where R and X are the resistive and reactive parts of the load, and Z_o is the transmission-line impedance.

References

[1] Jasik, *Antenna Engineering Handbook*, first edition, McGraw Hill, 1961, pp.3-2 through 3-7.
[2] *The Radio Amateur's Handbook*, fifty-eighth edition, ARRL, 1981, p 19-2.

The antenna had to be at least 3 feet in diameter to cover the whole 80-meter band

By Frank Witt, AI1H   From *QST*, April 1989

The Coaxial Resonator Match and the Broadband Dipole

It's easy to build a dipole with the coaxial resonator match. The SWR bandwidth of the antenna is almost triple that of a conventional dipole!

Out of the search for a simple dipole with acceptable SWR over the entire 80-meter band has come a matching technique with broadbanding properties and potential for many applications. This antenna and matching technique are extensions of work described by the author in October 1986 *QST*.[1] A review of that article is recommended as background.

In the sections that follow, the complete description of an 80-meter broadband dipole is provided, including performance data and construction details. Then, for those who want a better understanding of how it works, an explanation of the theory behind the broadband dipole and the coaxial resonator match is provided. Some other applications are described, one of which will be of great interest to 80-meter DX hunters.

An 80-Meter Broadband Dipole

Fig 1 shows the detailed dimensions of the 80-meter coaxial-resonator-match broadband dipole. Notice that the total length of the coax is an electrical quarter wavelength, has a short at one end, an open at the other end, a strategically placed crossover and is fed at a T junction. (The crossover is made by connecting the shield of one coax segment to the center conductor of the adjacent segment and by connecting the remaining center conductor and

shield in a similar way.) At AI1H, the antenna is constructed as an inverted-V dipole with a 110° included angle and an apex at 60 feet. The measured SWR v frequency is shown in Fig 2. Also in Fig 2 is the SWR characteristic for an uncompensated inverted-V dipole made from the same materials and positioned exactly as was the broadband version. SWR measurements were made with a Daiwa Model CN 520 cross-needle SWR/power meter. Corrections were made for the cable loss between the antenna and the meter.

The antenna, made from RG-8 coaxial cable and no. 14 AWG wire, is fed with 50-Ω coax. The coaxial cable should be cut

[1]Notes appear on page 1-22.

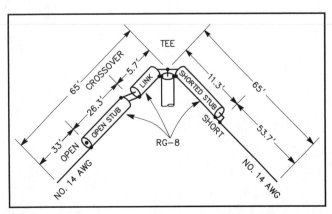

Fig 1–Coaxial-resonator-match broadband dipole for 80 meters. The coax segment lengths total ¼ wavelength. The overall antenna length is the same as that of a conventional inverted-V dipole.

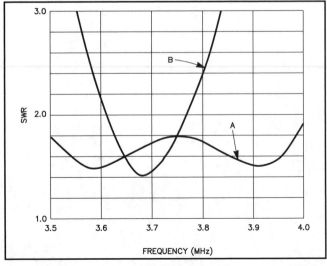

Fig 2–Curve A, the measured performance of the antenna of Fig 1. Also shown for comparison is the measured SWR of the same dipole without compenssation, curve B.

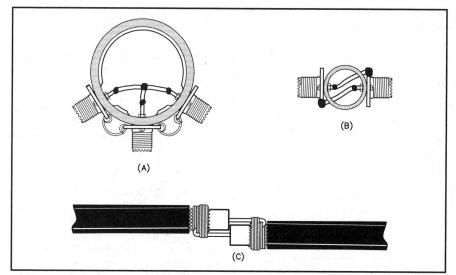

Fig 3—T and crossover construction. At A, a 2-inch PVC pipe coupling can be used for the T, and at B, a 1-inch coupling for the crossover. These sizes are the nominal inside diameters of PVC pipe that fits these couplings. The T could be standard UHF hardware (an M-358 T and a PL-258 coupler). An alternative construction for the crossover is shown at C, where a direct solder connection is made.

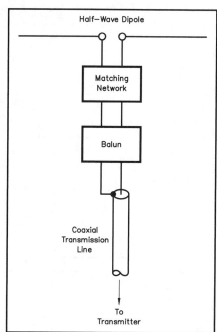

Fig 4—The broadband antenna system.

so the stub lengths of Fig 1 are within ¹/₂ inch of the specified values. PVC plastic pipe couplings and SO-239 UHF chassis connectors can be used to make the T and crossover connections, as shown in Fig 3 at A and B. Alternatively, a standard UHF T connector and coupler can be used for the T, and the crossover can be a soldered connection (Fig 3C). I used RG-8 coax because of its ready availability, physical strength, power handling capability and moderate loss. An RG-58 model was also designed and built, and it performed well electrically. I don't recommend the RG-58 version, however, because it is too fragile. For example, the coaxial cable stretches enough from its own weight to affect the tuning. Also, RG-58 will have substantially lower power-handling capability than RG-8.

Cut the wire ends of the dipole about three feet longer than the lengths given in Fig 1. If there is a tilt in the SWR v frequency curve when the antenna is first built (a lopsided "W" shape), it can be flattened to look like the shape of curve A in Fig 2 by increasing or decreasing the wire length. Each end should be lengthened or shortened by the same amount. Try 6-inch changes at each end with each adjustment. Increasing the dipole length will lower the SWR at the low end of the band; decreasing the dipole length will lower the SWR at the high end of the band.

A word of caution: If the chosen coaxial cable is not RG-8 or equivalent, the dimensions will have to be modified. For example, RG-8X has a different insulation material than RG-8, and its use would dictate different segment lengths. The following cable types have about the same characteristic impedance loss and velocity factor as RG-8 and could be substituted: RG-8A, RG-10, RG-10A, RG-213 and RG-215.

Important point: The calculated coaxial segment lengths were based on the assumption that the Q and radiation resistance at reso-

nance of the uncompensated dipole were 11.5 and 70 Ω), respectively. If the Q and radiation resistance differ markedly from these numbers because of different ground characteristics, antenna height, surrounding objects and so on, then different segment lengths would be required. In fact, if the dipole Q is too high, broadbanding is possible, but an SWR under 2:1 over the whole band cannot be achieved. More is said on the practical limitations of the coaxial resonator match in a later section of this article.

What is the performance of this broadband antenna relative to that of a conventional inverted-V dipole? Aside from the slight loss (about 1 dB at band edges, less elsewhere) because of the nonideal matching network, the broadband version behaves essentially the same as a dipole cut for the frequency of interest. That is, the radiation patterns for the two cases are virtually the same. In reality, the dipole itself is not "broadband,") but the coaxial resonator match provides a broadband match between the transmission line and the dipole antenna. This match is a remarkably simple way to broaden the SWR response of a dipole.

Broadbanding the Dipole

Fig 4 shows a broadband antenna system containing a coaxial (unbalanced) transmission line, a balun, a matching network and the half-wave dipole antenna. Use of the balun is recommended in order to

prevent radiation from the feed line.

The matching network is a transformer and a resonant circuit, as shown in Fig 5. Such an arrangement has been used in the past to achieve broadbanding.[2-4] The resonant circuit will have a finite Q, and this is the value of Q that determines the loss of efficiency caused by the matching network. The resonant circuit can be realized with LC components or with transmission-line segments. In fact, the transformer function can be performed with the same components. In the reference of note 1, it was shown how an LC resonator can act as the transformer as well. The transmission-line transformer can also be used to achieve the necessary impedance transformation, as is shown shortly.

The Coaxial Resonator Match

The coaxial resonator match performs the same functions as its predecessors, the T match and the gamma match, ie, that of matching a transmission line to a resonant dipole.[5] These familiar matching devices, as well as the coaxial resonator match, are shown in Fig 6. The coaxial resonator match has some similarity to the gamma match. It allows connection of the shield of the coaxial feed line to the center of the dipole and it feeds the dipole off-center, although center feed is also possible. The coaxial resonator match has a further advantage: It can be used to broadband the antenna system while it is providing

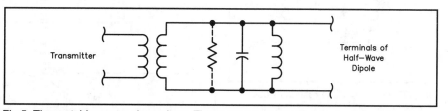

Fig 5—The matching network topology. The resonant circuit provides broadbanding by compensating for the reactance of the dipole, while the transformer adjusts the impedance level of the antenna feed to an optimum value.

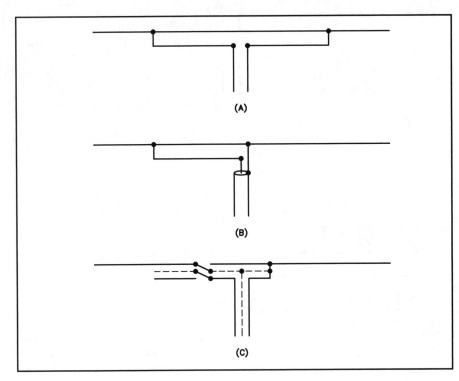

Fig 6—Dipole matching methods. At A, the T match; at B, the gamma match; at C, the coaxial resonator match.

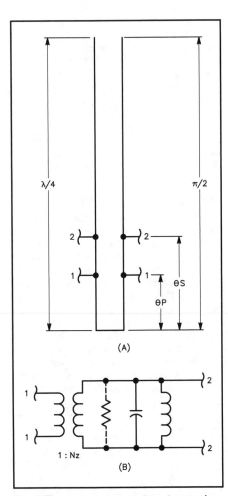

Fig 7—The quarter-wave resonator used as a transformer. Notice from the equivalent circuit of B that a simple piece of transmission line can provide the function of a tuned transformer.

an impedance match.

The coaxial resonator match is a resonant transformer made from a quarter wavelength piece of coaxial cable. It is based on a technique used at VHF and UHF to realize a low-loss impedance transformation.[6] Fig 7 shows how a quarter-wavelength transmission line with a short at one end and an open at the other end will provide transformer action over a limited band. Note that the equivalent circuit consists of a transformer with a parallel-tuned circuit connected across its secondary. The equivalent resonator has a Q, QN, which is related to the loss of the coax at the frequency of interest:

$$QN = \frac{2.774 \, F0}{A \times VF} \qquad (Eq \ 1)$$

where

F0 = resonant frequency (MHz)
A = resonator transmission-line attenuation (decibels/100 ft)
VF = velocity factor of resonator coax

The approximate impedance transformation ratio is given by

$$NZ = \frac{\sin^2 \theta S}{\sin^2 \theta P} \qquad (Eq \ 2)$$

where θS and θP are the electrical angles (lengths) of the secondary and primary taps, respectively, measured from the shorted end of the resonator.

For example, if the secondary tap were 0.1 wavelength from the short and the primary tap were 0.06 wavelength from the short, then

$$NZ = \frac{\sin^2 \left(2\pi \times 0.1\right)}{\sin^2 \left(2\pi \times 0.06\right)} = 2.5$$

For the practical application of matching the coaxial cable to the broadband dipole, the desired impedance transformation ratio can be readily obtained. The resonator transformer impedance transformation ratio is analogous to a conventional transformer, where

$$NZ = \frac{NS^2}{NP^2} \qquad (Eq \ 3)$$

where NS and NP are the number of secondary and primary turns, respectively. The resonator impedance level (impedance of resonator inductance or capacitance at resonance) is given by

$$ZN = \frac{4 \, ZR}{\pi} \sin^2 \theta S \qquad (Eq \ 4)$$

where ZR = characteristic impedance of line (ohms).

Off-Center Feed

The reason for the use of coaxial cable for the resonator will be seen later. But first, consider the concept of feeding the dipole off center. In most cases, half-wave dipoles are split and fed at the center. However, off-center feed is possible and has been used before. Two examples are the so-called Windom antenna and the dipole using the gamma match. Fig 8 shows a dipole with off-center feed. If you assume that the current distribution over the dipole is sinusoidal in shape, with zero current at the ends and maximum current at the center (and this is usually a very good assumption), then the radiation resistance at resonance is modified as follows.

$$RAF = \frac{RA}{\cos^2 \theta D} \qquad (Eq \ 5)$$

where
RAF = the radiation resistance at the feed point (ohms)
RA = the radiation resistance at the

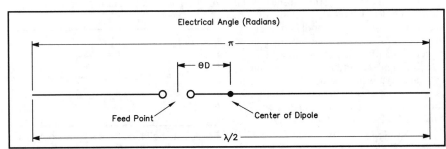

Fig 8—The dipole with off-center feed. See text regarding the feed-point impedance.

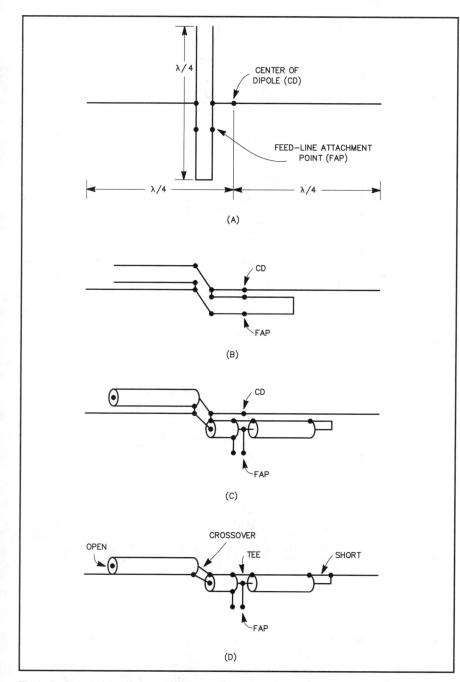

Fig 9—Evolution of the coaxial-resonator-match broadband dipole. At A, the resonant transformer is used to match the feed line to the off-center-fed dipole. The match and dipole are made collinear at B. At C, the balanced transmission-line resonator/transformer of A and B is replaced by a coaxial version. Since the shield of the coax can serve as part of the dipole radiator, the wire adjacent to the coax match can be eliminated, D.

center of the dipole (ohms)

θD = the electrical angular distance off of center

For example, if the radiation resistance of the dipole at its center were 72 Ω, then, if it were fed off center by 0.03 wavelength (θD = $2\pi \times 0.03$ = 0.188 radians), the radiation resistance at the feed point would be

$$RAF = \frac{72}{\cos^2 0.188} = 74.6 \text{ ohms}$$

In the practical cases I considered, the change in antenna feed-point impedance arising from off-center feed is small, but it should be taken into account for best results.

The Coaxial Resonator Matched Broadband Dipole

All of the necessary elements of the broadband dipole have now been described. It remains to assemble them into an antenna system. If you compare Figs 5 and 7, you can see that the coaxial resonator match contains the necessary elements for matching and broadbanding. The off-center feed concept provides the finishing touch.

Fig 9 shows the evolution of the broadband dipole. Now it becomes clear why coaxial cable is used for the quarter-wave resonator/transformer; interaction between the dipole and the matching network is minimized. The effective dipole feed point is

located at the crossover. In effect the match is physically located "inside" the dipole. Currents flowing on the inside of the shield of the coax are associated with the resonator, currents flowing on the outside of the shield of the coax are the usual dipole currents. Skin effect provides a degree of isolation and allows the coax to perform its dual function. The wire extensions at each end make up the remainder of the dipole, making the overall length equal to a half wavelength.

The coaxial resonator match, like the gamma match, allows you to connect the shield of the coaxial feed line to the center of the dipole. If the antenna were completely symmetrical, then the RF voltage would be zero (relative to ground) at the center and no balun would be required. In the real situation, some voltage (again referred to ground) does exist at the dipole midpoint (as it does with the gamma match), but in many practical cases no balun is required. If one is used, it should be of the "choke" or "current balun" variety.[7-8] A longitudinal choke can be made by threading several turns of coax through a ferrite toroid, or a commercial variety, such as the W2DU balun, is appropriate. I've used the coaxial resonator matched broadband dipole both with and without a balun with little difference in SWR characteristic. However, there are situations where the balun would be required. To be safe, use a balun.

A useful feature of an antenna using the coaxial resonator match is that the entire antenna is at the same dc potential as the feed-line potential, thereby avoiding charge buildup on the antenna. Hence, noise and the potential of lightning damage are reduced.

Other Applications

The design of Fig 1 can be modified to yield an "80-Meter DX Special." In this case the band extends from 3.5 MHz to 3.85 MHz. Over that band the SWR is better than 1.6:1 and the calculated matching network loss is less than 0.75 dB. See Fig 10 for measured performance of an 80-Meter DX Special built and used by Ed Parsons, K1TR.

Design dimensions for the 80-Meter DX Special are given in Fig 11. The coax segment lengths are based on the assumption that the dipole Q and radiation resistance at resonance are 13 and 60 ohms, respectively. The calculated SWR for the uncompensated dipole and the coaxial resonator matched broadband dipole are shown in Fig 12. Since most amateurs do not know what Q and radiation resistance would exist for their installation, it is desirable to know how sensitive the SWR characteristic is to those parameters. With the aid of a simulation program, a deviation study was made for Q over the range 10 to 16 and radiation resistance ranging from 50 to 70 ohms. In the analysis, the coax segment lengths were not changed from the values shown in Fig 11. The results, given in Fig 13, show that the coaxial-resonator matched dipole is very robust. The SWR is less than 2:1 over virtually the entire 3.5- to 3.85-MHz band for the wide range of Q and radiation resistance values simulated. An obvious applica-

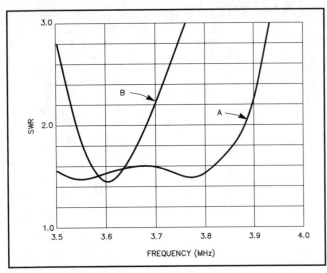

Fig 10—Measured SWR performance of the 80-Meter DX Special, curve A. Note the substantial broadbanding relative to a conventional uncompensated dipole, curve B.

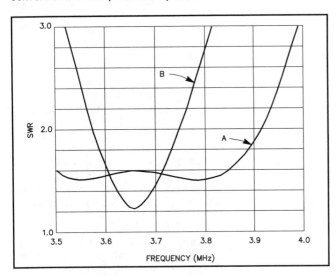

Fig 12—Calculated SWR response of the 80-Meter DX Special, curve A, and a conventional dipole, curve B.

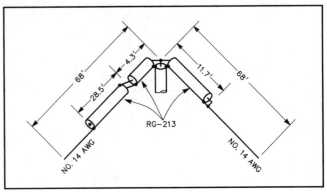

Fig 11—Dimensions of the 80-Meter DX Special, an antenna optimized for the phone and CW DX portions of the 80-meter band. Also see Fig 12.

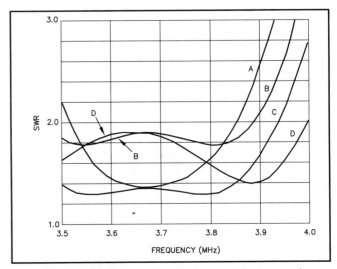

Fig 13—The results of a deviation study reveal the expected performance of the antenna of Fig 11 for a variety of conditions. The various curves were obtained with these parameters. Curve A: Q=16, R=70 Ω. Curve B: Q=16, R=50 Ω. Curve C: Q=10, R=70 Ω. Curve D: Q=10, R=50 Ω.

tion of the coaxial resonator match is to broadband a 160-meter dipole to cover the entire 1.8- to 2.0-MHz band. A design similar to the 80-meter antenna described in this article would have an SWR better than 1.5:1 over the whole 160-meter band. The calculated matching network loss is less than 1.1 dB.

The same concept might be applied to broadband a Yagi array, where you must usually settle for a compromise among gain, front-to-back ratio and SWR bandwidth. By applying the coaxial-resonator-matching principle, the SWR bandwidth of the array might be made wide enough that the gain and front-to-back ratio could be better optimized. This conjecture was suggested by John Kenny, W1RR.

The coaxial resonator match can be applied to monopoles as well. In this case, one of the coax segments could be located "inside" one of the radials. Other applications include full-wave loops and 3/2-wavelength center-fed antennas.

Summary and Acknowledgments

The coaxial resonator match is a form of matching network for use between the transmission line and the antenna. This match, which becomes an integral part of a dipole antenna, serves not only as a matching device, but also has inherent broadbanding properties. The 80-meter broadband dipoles presented as practical examples ably demonstrate the utility of the coaxial resonator match. This antenna achieves a long-sought-after goal of realizing a simple dipole that is well matched over the entire 80-meter band.

I am indebted to my wife, Barbara, N1DIS, for her encouragement throughout the course of this project. Also, several discussions with John Kenny, W1RR, provided inspiration during the course of the development of the coaxial resonator match. Further, an example of a broadband dipole shown to me by Reed Fisher, W2CQH, initiated my investigation which led to the relatively efficient design presented in this paper. I am grateful to Ed Parsons, K1TR, who first constructed and evaluated the 80-Meter DX Special.

A complete description of how one can use the coaxial resonator match in other applications is in *The ARRL Antenna Compendium, Volume 2*. That paper also contains design equations for computing the segment lengths of the coaxial resonator match.

Notes
[1] F. J. Witt "Broadband Dipoles—Some New Insights," *QST*, Oct 1986, pp 27-37.
[2] J. Hall, "The Search for a Simple, Broadband 80-Meter Dipole," *QST*, Apr 1983, pp 22-27.
[3] R. D. Snyder, "Broadband Antennae Employing Coaxial Transmission Line Sections," United States Patent no. 4,479,130, issued Oct 23, 1984.
[4] R. C. Hansen, "Evaluation of the Snyder Dipole," *IEEE Transactions on Antennas and Propagation*, Vol. AP-35, No. 2, Feb 1987, pp 207-210.
[5] G. L. Hall, Ed, *The ARRL Antenna Book* (Newington: ARRL); 14th ed (1982), pp 5-13–5-14; 15th ed (1988), pp 26-16–26-18.
[6] *The Radio Amateur's Handbook*, 52nd ed (Newington: ARRL, 1975) pp 54-55.
[7] M. W. Maxwell, "Some Aspects of the Balun Problem," *QST*, Mar 1983, pp 38-40.
[8] R W. Lewallen, "Baluns: What They Do and How They Do It," *The ARRL Antenna Compendium*, Vol. 1 (Newington: ARRL, 1985), pp 157-164.

By Rudy Severns, N6LF

A Wideband 80-Meter Dipole

This worthy antenna is so simple and inexpensive you'll want more than one!

The 500-kHz width of the 80-meter band makes it by far the widest HF amateur band on a percentage basis—13% of the center frequency. Over the years, a legion of articles have described antennas that purported to provide an SWR of less than 2:1 over the whole band. Some did, some didn't. With my two transceivers—a Drake TR-7 and Yaesu FT-757GX—even a 2:1 SWR isn't low enough because the rigs automatically begin to reduce output power before a 2:1 SWR is reached. I suspect this is not an uncommon occurrence with other rigs (not equipped with built-in automatic antenna tuners) as well.

What's really needed is an antenna that provides an SWR below 1.6 or 1.7:1 over the entire band. It'd be really convenient to jump from one end of the band to the other without having to think about retuning the antenna tuner or rig, or buying an automatic antenna tuner. Such a requirement makes antenna design tough!

The following is a description of an antenna that meets the need. This one has been built—and it works great with no noticeable SWR degradation caused by rain, snow, wind or other weather elements. Surprisingly, it's a simple wire antenna that's only as long as a standard dipole.

Earlier Antennas

My idea has its roots in two well-known antennas: the open-sleeve dipole[1,2,3] and the

[1]Notes appear on page 1-25.

The added conductors create new resonances. This effect can be used to multiband or broadband an antenna.

folded dipole.[4] With an open-sleeve dipole, additional conductors are added in close proximity to—but not connected to—a common single-wire dipole, as shown in Figure 1. In addition to the fundamental resonance of the simple dipole, the added

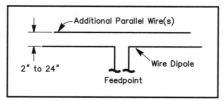

Figure 1—An open-sleeve dipole example.

conductors create new resonances. This effect can be used to multiband or broadband an antenna—and it's an idea that's been around since WW II.

A folded dipole's bandwidth is greater than a single-wire dipole made of the same wire size. Although the bandwidth attainable with a folded dipole is better, by itself, it's still not good enough for our needs. Figure 2 shows the typical SWR plot for a folded dipole, using 12-inch element spacing, #14 wire and centered on 3.750 MHz. This antenna's 2:1 SWR bandwidth is approximately 375 kHz. You can improve things a bit by using greater element spacing, but then the weight and length of the

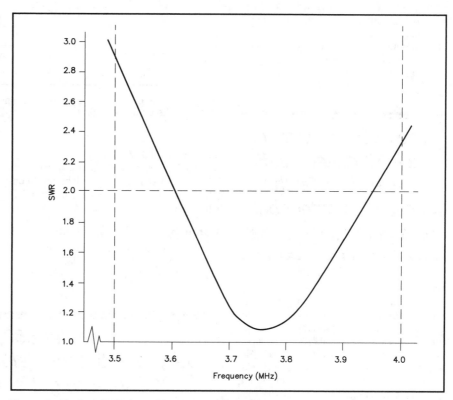

Figure 2—Typical SWR for a folded dipole resonant at 3.75 MHz.

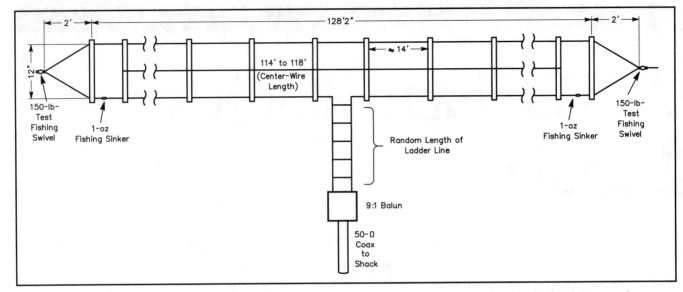

Figure 3—The open-sleeve folded dipole: simple and inexpensive. The antenna is fed with a random length of 450-Ω open-wire transmission line through a 9:1 balun.

spacers gets to be a hassle and you still won't have sufficient bandwidth.

Antenna Height

One common problem with any 80-meter antenna is installing it high enough to do some good. Because the current maximum is at the center of the dipole, it's important to keep that part of the antenna as high as possible. For most installations, 70 feet is pretty high, but at 80 meters, 70 feet is only one quarter of a wavelength.

A dipole's radiation angle is largely determined by the height of its center. If the antenna is strung between two supports, there's bound to be some sag, height is lost and the radiation angle raised. Weight of any sort—baluns, long lengths of coax, matching networks, etc, particularly near the antenna's center—contribute to sag. The resultant high-angle radiation is great for local QSOs, but bad news for DXing.

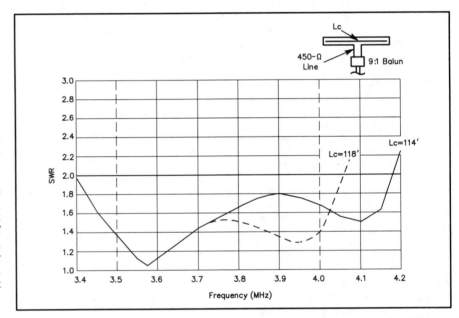

Figure 4—SWR curve of the open-sleeve dipole of Figure 3, showing curves for different lengths of the center wire (L_C).

The 500-kHz width of the 80-meter band makes it by far the widest HF amateur band on a percentage basis.

If I can't provide support for the antenna center, I prefer to use lightweight transmitting twinlead (weighing in at 2.4 lb/100 ft versus the 9.4 lb/100 ft of RG-8 coax), with the balun at ground level. The 450-Ω ladder line is quite efficient, relatively light and costs about 16 cents a foot, much less than coax.

The feed-point impedance of a folded di-

pole is about 300 Ω. Although 300-Ω ladder line is available, making the transition from 300 Ω to 50 Ω requires a 6:1 balun. Such baluns can be bought or made, but 4:1 or 9:1 baluns are much more common.

A Broadband 80-Meter Antenna

The antenna I came up with is shown in Figure 3. It's simply an open-sleeve version of the folded dipole. The resonator wire added

midway between the two folded dipole elements is supported by the spacers already used for the folded dipole. That's all there is to it: a single wire down the middle of a folded dipole! One interesting result of adding the wire is that not only is the antenna very broadband, but by juggling the spacing and wire lengths a bit, Z_r is very close to 450 Ω, which fits in nicely with available transmission lines and a 9:1 balun. The transmission line oper-

---

So, for $50, you've got everything but the balun and the lead-in coax. You've also got a darn good antenna.

---

ates with a very low SWR and can be of virtually any length.

A graph of the measured SWR for two lengths of the center wire (L_c) is shown in Figure 4. The measurements were made with considerable care, using Bird wattmeters. For L_c = 118 feet, the highest SWR is 1.5:1, and is less than that over most of the band. For L_c = 114 feet, the worst-case SWR is 1.8:1, but the overall 2:1 bandwidth is extended to 800 kHz. This would be advantageous to MARS operators operating just above the upper band edge. Experimenting further, I shortened L_c to 112 feet, which pushed the 2:1 bandwidth up to nearly 1 MHz (3.3 to 4.25 MHz). For most hams, that may not be of great importance, but it's something to keep in mind.

Figure 3 shows the number and separation of the wire spacers. It's important to keep the spacers as light (and inexpensive!) as possible. The two spacers on each end have to be fairly stiff, so I used sections cut from solid fiberglass electric-fence wands.[5] The rest of the spreaders are made from half sections of $^1/_2$-inch CPVC plastic pipe. They're about half the weight of the fiberglass wand spreaders. I could have used full sections of the CPVC pipe for the end spreaders but, for the same weight, they would have had more wind loading.

Summary

Modeling this antenna, which is essentially a transmission line, doesn't work very well on *MININEC*-based programs.[6] *NEC* programs such as *NecWires*[7] are needed, and even then, you have to use 50 to 100 segments per $\lambda/2$ for the final design. Using *NecWires*, the total computed loss was only

That's all there is to it: a single wire down the middle of a folded dipole!

0.07 dB (1.6%) for #14 wire and 0.09 dB (2%) for #16. Combined with the very low loss of the open-wire transmission line, if a good three-core, 9:1 Guanella balun[8] is used, the overall efficiency will be quite good.

At best, these antennas will be close to the ground in terms of wavelengths. The ground effects are important and will affect the impedances and final dimensions. This antenna was modeled at a height of 70 feet over poor ground (ε_r = 13, σ = 2 mS), which corresponds (more or less) to my location and support height. I only had to adjust the center wire a bit to get the predicted performance. At another location or antenna height, the final performance and dimensions may be different.

A folded dipole *loves* to rotate when being hoisted and it twists when the wind blows, which really upsets the SWR if the parallel wires short together. In Figure 3, I've included a couple of details that help reduce this problem. The ends of the dipole are not symmetrical. To aid in avoiding antenna twist, 1-oz fishing sinkers are added to the bottom wire on each end. I also use two heavy-duty (150-

pound-capacity) fishing-line swivels at the antenna support points.

Performance isn't the only criterion for a good antenna. For most hams, cost is always a consideration. This design uses 380 feet of wire (a total cost of $34 at 9 cents per foot[9]) and about a buck's worth of $^1/_2$-inch CPVC pipe. The CPVC can also be used for the center and end insulators. The 450-Ω open-wire transmission line costs 14 to 16 cents per foot (see Note 9), so add another $15 to the total. So, for $50, you've got everything but the balun and the lead-in coax. You've also got a darn good antenna.

Notes

[1]Roger Cox, WBØDGF, "The Open-Sleeve Antenna: Development of the Open-Sleeve Dipole and Open-Sleeve Monopole for H.F. and V.H.F. Amateur Applications," *CQ*, Aug 1983, pp 13-19.

[2]Gary Breed, K9AY, "Multi-Frequency Antenna Technique Uses Closely Coupled Resonators," *RF Design*, Nov 1994, pp 78-85.

[3]Bill Orr, W6SAI, "Radio FUNdamentals," The Open Sleeve Dipole, *CQ*, Feb 1995, pp 94-98.

[4]R. Dean Straw, N6BV, Ed, *The ARRL Antenna Book* (Newington: ARRL, 17th ed, 1994), p 2-32.

[5]These wands, which measure $^3/_8$-inch in diameter and are 4 feet long, are available from farming supply stores and Sears.

[6]*ELNEC* is available from Roy Lewallen, W7EL, PO Box 6658, Beaverton, OR 97007. *AO 6.1* is available from Brian Beezley, K6STI, 3532 Linda Vista Dr, San Marcos, CA 92069, tel 619-599-4962.

[7]*NEC-Wires* is available from Brian Beezley, K6STI; see Note 6.

[8]Jerry Sevick, W2FMI, *Transmission Line Transformers* (Newington: ARRL, 2nd ed, 1990), p 9-28.

[9]Several *QST* advertisers offer 450-Ω open-wire transmission line. See the ads at the back of any issue.

CHAPTER TWO
MULTIBAND DIPOLES

By H. J. Berg, W3KPO

From *QST*, July 1956

Multiband Operation with Paralleled Dipoles

A simple antenna system with coax feed

The increased flexibility of the modern amateur transmitter accents the need for a single antenna for multiband operation. Few have space for an antenna farm; most of us must be content with the average-sized city lot. As a result, we can erect but one antenna.

There are several well-known ways of achieving multiband operation with one antenna. It is most often accomplished by the use of tuned feeders, or a long wire fed at the end without feeders. This requires an antenna tuner. To change bands, it is necessary to change coils in the tuner, rearrange taps, and retune. On the other hand, if the single antenna is fed with a nonresonant line, its operation must be limited to the one band for which it is cut.

An antenna commonly referred to as the "300-ohm off-center-fed Windom," has gained considerable popularity, because it is reputed to be a system that permits operation on several bands with a "flat" line. However, those who have investigated the design of this antenna have discovered that its practical operation must involve considerable compromise.[1] Many more have learned by experience that there are other disadvantages. Unbalanced feeder currents result in feeder radiation. Radiation from the transmission line as well as from the antenna gives rise to variation in the radiation pattern from band to band, feedpoint impedance variation and the appearance of r.f. on supposedly grounded circuits and house wiring. The last-mentioned effect has been so bad in some cases as to ruin the performance of a well-shielded v.f.o. In spite of these troubles, many have continued to use it from lack of any other choice.

For some time, the author, in common with many others, has been in search of a simple multiband system that could be fed efficiently with coax cable. Finally, an item that appeared in *Radiotron Designer's Handbook*[2] served as a reminder of a system that was described in *QST* many years ago. The arrangement is so simple that it is

surprising that more frequent use has not been made of it in recent years.

The principle of the system is shown in Fig. 1. The arrangement consists of separate dipoles for each band, all connected in parallel to a single coaxial transmission line. With one of the dipoles operating at its resonant frequency, its feedpoint impedance will, of course, be suitable for matching a low-impedance line (approximately 70 ohms). The remaining *lower-frequency* dipoles will be at or close to harmonic resonance at the operating frequency. However, since their halves

> The arrangement consists of separate dipoles for each band, all connected in parallel to a single coaxial transmission line.

will be in phase, the impedance presented to the line will be high and essentially resistive. This high impedance will be in parallel with the 70 ohms of the active dipole and therefore will have negligible effect on the line termination, and little current will flow to the longer dipoles.

The remaining *higher-frequency* dipoles will present an impedance consisting of resistance and capacitive reactance. However, the resulting impedance will also be high compared to the 70 ohms of the active dipole.

A departure occurs in the case of 15-meter operation. On this band, the 7-MHz antenna will be close to resonance at its third harmonic, and its center impedance will be low—of the order of 100 ohms or so. Also, at this frequency (21 MHz), the 14-MHz dipole will show inductive reactance, while the 28-MHz dipole will show capacitive reactance. However, the resultant is still high compared to 70 ohms. Since the 7 MHz dipole presents a fairly close match to the line, considerable power will be fed to it, and there is some question as to the value of including a separate dipole for the 15-meter band. The radiation pattern of the 7-MHz dipole at its third harmonic will be essentially nondirective, although there will be fairly sharp nulls at

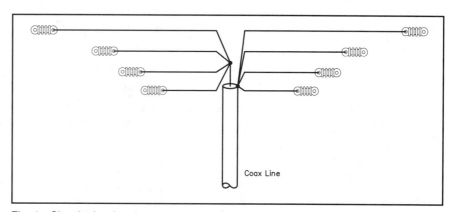

Fig. 1—Sketch showing the arrangement of parallel half-wave antennas in the coax-fed multiband antenna system.

angles of about 20 degrees with the direction of the dipole wire.

Construction

There are several ways in which the antenna elements can be suspended. W9YPQ suspended one set of elements from the one above it, using insulator-terminated wood spreaders about one foot long. In the author's installation, the elements are simply allowed to droop about two feet, one below the other. Ropes attached to the end insulators are brought back up to a common anchoring point. The elements could also be fanned out, either vertically or horizontally, provided that the angle of fanning does not become too great. The lengths of the elements should, of course, be the same as those for dipoles for each band.

Several other local hams are using this antenna system with excellent results, and much DX has been worked. With a bandswitching transmitter we can hop from band to band as quickly as with the receiver, since there is no need to fuss around with an antenna tuner.[3] Checks with a Millen s.w.r. bridge show that the s.w.r. never exceeds 1.5 to 1[4] on any of the bands on which the antenna has been operated. These include 40, 20, 15 and 10 meters. Those who have space can add an 80-meter dipole, of course.

Notes
[1] Wrigley, "Characteristics of Harmonic Antennas," QST, February, 1954.
[2] RCA, Harrison, NJ.
[3] This system will, of course, respond to harmonics and submultiples of the output frequency. Therefore more than ordinary care must be used in suppressing these frequencies. —Ed.
[4] Confirming by measurements on a similar system at ARRL HQ—Ed.

From QST, August 1968 (Technical Correspondence)

Multiband Antenna

Technical Editor, QST:

The antenna setup shown in Fig. 1 is the result of an attempt to find a system that would be matched well enough on all bands so that the load presented to the transmitter would be within the limits that normal equipment can handle. I started from the Collins broad-band antenna, which used fanned dipoles spread 12 inches at the ends, and tried to treat them as parallel dipoles by cutting one shorter than the other—net result failure. While the Collins configuration was broader it was not broad enough. I finally found that when parallel dipoles near the same frequency were fanned over 12 feet they started to act independently. The present system consists of a balun suspended by glassline between two masts 40 feet apart and fed with RG-8A/U. From this point I run a dipole cut for 3975 and one cut for 3600, spread 40 feet at the ends. Between these points I also run one cut for 7050 and one cut for 7250. In the center I have one cut for 14,250.

My s.w.r. remains well below 1.5:1 over the complete 75/80-, 40-, and 20-meter bands. It is less than 2:1 over 15, with less than 1.5:1 over all the phone section and about half the c.w. portion. The system will also work on 10, with less than 2:1 over most of the band but no actual indication of resonance.

More wires could be used on any band, as all elements here are in inverted-V form and some can be suspended under others. I

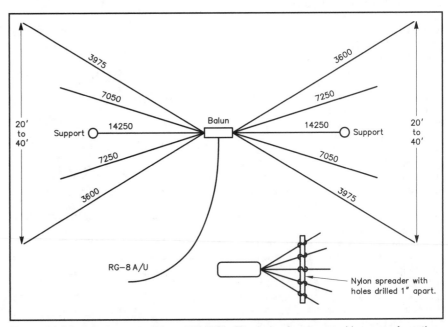

Fig. 1—Multiband antenna system at W4RZL. The balun is supported by ropes from the supports; the 14-MHz antenna is fastened to it in line with the rope but is not part of the supporting structure. The 7-MHz antenna can droop directly under those for 3.5-4 MHz. Small drawing shows method of fastening antenna wires at the balun to prevent twisting.

have even used a 75-meter wire in the center of the band and achieved a flat line over the entire band, as close as I could determine with the Collins wattmeter.

All antennas were first cut long and then pruned with an Allied Radio bridge and a Millen dipper beat to the station receiver and checked with the wattmeter. Various lengths of coax—12, 25, 50 and 100 feet— have been spliced into the feed line and cause no significant change when inserted.— Howard L. Schonher, W4RZL

By Louis Varney, G5RV

From *The ARRL Antenna Compendium, Vol 1*

The G5RV Multiband Antenna . . . Up-to-Date

Adapted from an article of the same title in *Radio Communication*, July 1984, pp. 572-575.

The G5RV antenna, with its special feeder arrangement, is a multiband center-fed antenna capable of efficient operation on all HF bands from 3.5 to 28 MHz. Its dimensions are specifically designed so it can be installed in areas of limited space, but which can accommodate a reasonably straight run of about 102 ft for the flat-top. Because the most useful radiation from a horizontal or inverted-V resonant antenna takes place from the center two-thirds of its total length, up to one-sixth of this total length at each end of the antenna may be dropped vertically, semi-vertically, or bent at a convenient angle to the main body of the antenna without significant loss of effective radiation efficiency. For installation in very limited areas, the dimensions of both the flat-top and the matching section can be divided by a factor of two to form the half-size G5RV, which is an efficient antenna from 7 to 28 MHz. The full-size G5RV will also function on the 1.8-MHz band if the station end of the feeder (either balanced or coaxial type) is strapped and fed by a suitable matching network using a good earth connection or a counterpoise

> The G5RV antenna, with its special feeder arrangement, is a multiband center-fed antenna capable of efficient operation on all HF bands from 3.5 to 28 MHz.

wire. Similarly, the half-size version may be used on the 3.5- and 1.8-MHz bands.

In contradistinction to multiband antennas in general, the full-size G5RV antenna was not designed as a $\lambda/2$ dipole on the lowest frequency of operation, but as a $3\lambda/2$ center-fed long-wire antenna on 14 MHz, where the 34 ft open-wire matching section functions as a 1:1 impedance transformer. This enables the 75-ohm twin lead, or 50/80-ohm coaxial cable feeder, to see a close impedance match on that band with a consequently low SWR on the feeder. However, on all the other HF bands, the function of this section is to act as a "make-up" section to accommodate that part of the standing wave (current and voltage components) which, on certain operating frequencies, cannot be completely accommodated on the flat-top (or inverted-V) radiating portion. The design center frequency of the full-size version is 14.150 MHz, and the dimension of 102 ft is derived from the formula for long-wire antennas which is:

$$\text{Length (ft)} = \frac{492\,(n - 0.05)}{f_{MHz}}$$

$$= \frac{492 \times 2.95}{14.15}$$

$$= 102.57 \text{ ft (31.27 m)}$$

where n = the number of half wavelengths of the wire (flat-top).

Because the whole system will be brought to resonance by the use of a matching network in practice, the antenna is cut to 102 ft.

As the antenna does not make use of traps or ferrite beads, the dipole portion becomes progressively longer in electrical length with increasing frequency. This effect confers certain advantages over a trap

or ferrite-bead loaded dipole because, with increasing electrical length, the major lobes of the vertical component of the polar diagram tend to be lowered as the operating frequency is increased. Thus, from 14 MHz up, most of the energy radiated in the vertical plane is at angles suitable for working DX. Furthermore, the polar diagram changes with increasing frequency from a typical $\lambda/2$ dipole pattern at 3.5 MHz and a two $\lambda/2$ in-phase pattern at 7 and 10 MHz to that of a long-wire antenna at 14, 18, 21, 24 and 28 MHz.

Although the impedance match for 75-ohm twin lead or 80-ohm coaxial cable at the base of the matching section is good on 14 MHz, and even the use of 50-ohm coaxial cable results in only about a 1.8:1 SWR on this band, the use of a suitable matching network is necessary on all the other HF bands. This is because the antenna plus the matching section will present a *reactive* load to the feeder on those bands. Thus, the use of the correct type of matching network is essential in order to ensure the maximum transfer of power to the antenna from a typical transceiver having a 50-ohm coaxial (unbalanced) output. This means unbalanced input to balanced output if twin-wire feeder is used, or unbalanced to unbalanced if coaxial feeder is used. A matching network is also employed to satisfy the stringent load conditions demanded by such modern equipment that has an automatic level control system. The system senses the SWR condition present at the solid state transmitter output stage to protect it from damage, which could be caused by a reactive load having an SWR of more than about 2:1.[1]

The above reasoning does not apply to the use of the full-size G5RV antenna on 1.8 MHz, or to the use of the half-size version on 3.5 and 1.8 MHz. In these cases, the station end of the feeder conductors should be "strapped" and the system tuned

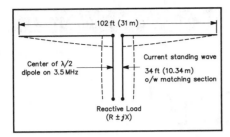

Fig 1—Current standing-wave distribution on the G5RV antenna and matching section at 3.5 MHz. The antenna functions as a λ/2 dipole partially folded up at the center.

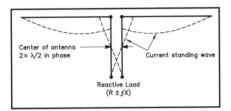

Fig 2—Current distribution on the antenna and matching section at 7 MHz. The antenna now functions as a col-linear array with two half waves fed in phase.

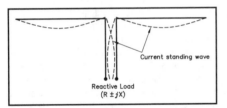

Fig 3—Current standing-wave distri-bution on the antenna and matching section at 10 MHz. The antenna functions as a collinear array with two half waves fed in phase.

to resonance by a suitable series-connected inductance and capacitance circuit connected to a good earth or counterpoise wire. Alternatively, an unbalanced-to-unbalanced type of matching network such as a T or L matching circuit can be used.[2] Under these conditions the flat-top (or inverted-V) portion of the antenna, plus the matching section and feeder, function as a Marconi or T antenna, with most of the effective radiation taking place from the vertical, or near vertical, portion of the system; the flattop acts as a top-capacitance loading element. However, with the system fed as described above, very effective radiation on these two bands is obtainable even when the flat-top is as low as 25 ft above ground.

Theory of Operation

The general theory of operation has been explained above. The detailed theory of operation on each band from 3.5 to 28 MHz follows, aided by figures showing the current standing wave conditions on the flat-top, and the matching (or make-up) section. The relevant theoretical horizontal plane polar diagrams for each band may be found in any of the specialized antenna handbooks. However, it must be borne in

mind that: (a) the polar diagrams generally shown in two dimensional form are, in fact, three dimensional (i.e., solid) figures around the plane of the antenna; and (b) all theoretical polar diagrams are modified by reflection and absorption effects of nearby conducting objects such as wire fences, metal house guttering, electric wiring systems, and even large trees. Also, the local earth conductivity will materially affect the actual polar radiation pattern produced by an antenna. Theoretical polar diagrams are based on the assumptions that an antenna is supported in "free space" above a perfectly conducting ground. Such conditions are obviously impossible of attainment in the case of typical amateur installations. What this means in practice is that the reader should not be surprised if *any* particular antenna in a typical amateur location produces contacts in directions where a null is indicated in the theoretical polar diagram, and perhaps not such effective radiation in the directions of the major lobes as theory would indicate.

3.5 MHz: On this band each half of the flat-top, plus about 17 ft of each leg of the matching section, forms a foreshortened or slightly folded up λ/2 dipole. The remainder of the matching section acts as an unwanted, but unavoidable reactance between the *electrical* center of the dipole and the feeder to the matching network. The polar diagram is effectively that of a λ/2 antenna. See Fig. 1.

7 MHz: The flat-top, plus 16 ft of the matching section, now functions as a partially folded up two half waves in phase antenna producing a polar diagram with a somewhat sharper lobe pattern than a λ/2 dipole because of its collinear characteristics. Again, the matching to a 75 ohm twin-lead or 50/80 ohm coaxial feeder at the base of the matching section is degraded somewhat by the unwanted reactance of the lower half of the matching

section, but, despite this, by using a suitable matching network, the system loads well and radiates very effectively on this band. See Fig. 2.

10 MHz: On this band the antenna functions as a two half-wave in-phase collinear array, producing a polar diagram virtually the same as on 7 MHz. A reactive load is presented to the feeder at the base of the matching section but, as for 7 MHz, the performance is very effective. See Fig. 3.

14 MHz: At this frequency the conditions are ideal. The flat-top forms a 3 λ/2 long center-fed antenna which produces a multilobe polar diagram with most of its radiated energy in the vertical plane at an angle of about 14 degrees, which is effective for working DX. Since the radiation resistance at the center of a 3 λ/2 long-wire antenna supported at a height of λ/2 above ground of average conductivity is about 90 ohms, and the 34-ft matching section now functions as a 1:1 impedance transformer, a feeder of anything between 75 and 80 ohms characteristic impedance will see a nonreactive (i.e., resistive) load of about this value at the base of the matching section, so that the SWR on the feeder will be near 1:1. Even the use of 50 ohm coaxial feeder will result in an SWR of only about 1.8:1. It is assumed here that 34 ft is a reasonable average antenna height in amateur installations. See Fig. 4.

18 MHz: The antenna functions as two full-wave antennas fed in phase; it combines the broadside gain of a two-element collinear array with a somewhat lower zenithal angle radiation than a λ/2 dipole because of its long-wire characteristic. See Fig. 5.

21 MHz: On this band the antenna works as a long wire of five halfwaves, producing a multilobe polar diagram with effective low zenithal angle radiation. Although a high resistive load is presented to the feeder at the base of the make-up section, the system loads well when used in

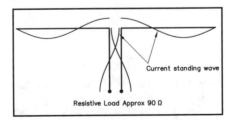

Fig 4–Current standing-wave distribution on the antenna and matching section at 14 MHz. In this case the antenna functions as a center-fed long wire of three half waves out of phase. The matching section now functions as a 1:1 impedance transformer, presenting a resistive load of approximately 90 ohms at the lower end.

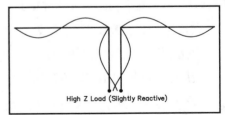

Fig 5–Current standing-wave distribution on the antenna and matching section at 18 MHz. The antenna functions as two full-wave antennas, slightly folded up at the center, fed in phase.

The local earth conductivity will materially affect the actual polar radiation pattern produced by an antenna.

conjunction with a suitable matching network and radiates effectively for DX contacts. See Fig. 6.

24 MHz: The antenna again functions effectively as a 5 λ/2 long wire, but because of the shift in the positions of the current antipodes on the flat-top and the matching section (Fig. 7), the matching or make-up section now presents a much lower resistive load condition to the feeder connected

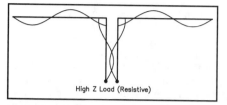

Fig 6–Current standing-wave distribution on the antenna and matching section at 21 MHz. On this band the antenna works as a long wire of five half waves. The base of the matching section presents a virtually nonreactive high impedance load to the feeder.

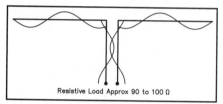

Fig 7–Current standing-wave distribution on the antenna and matching section at 24 MHz. The antenna functions as a long wire of five half waves.

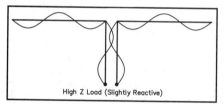

Fig 8–Current standing-wave distribution on the antenna and matching section at 28 MHz. The antenna functions as two long-wire antennas each of three half waves in length, fed in phase. A very effective form of antenna giving good multilobe, low zenithal angle, radiation.

to its lower end than it does on 21 MHz. Again, the polar diagram is multilobed with low zenithal angle radiation.

28 MHz: On this band, the antenna functions as two long-wire antennas, each of three half waves, fed in phase. The polar diagram is similar to that of a 3 λ/2 long-wire, but with even more gain over a λ/2 dipole because of the collinear effect obtained by feeding two 3 λ/2 antennas, in line and in close proximity, in phase. See Fig. 8.

Construction

The Antenna

The dimensions of the antenna and its matching section are shown in Fig. 9. If possible, the flattop should be horizontal and run in a straight line, and should be erected as high as can be above ground. In describing the theory of operation, it has been assumed that it is generally possible to erect the antenna at an average height of about 34 ft, which happens to be the optimum height for the antenna at 14 MHz. Although this is too low for optimum radiation ef-

As the antenna does not make use of traps or ferrite beads, the dipole portion becomes progressively longer in electrical length with increasing frequency. This effect confers certain advantages over a trap or ferrite-bead loaded dipole.

ficiency on 1.8, 3.5, and 7 MHz for any horizontal type of antenna, in practice few amateurs can install masts of the optimum height of half a wavelength at 3.5 or 7 MHz, and certainly not at 1.8 MHz.

If it is not possible to accommodate the 102-ft top in a straight line because of space limitations, up to about 10 ft of the antenna wire at each end may be allowed to hang vertically or at some convenient angle, or be bent in the horizontal plane, with little practical effect on performance. This is because, for any resonant dipole antenna, most of the effective radiation takes place from the center two-thirds of its length where the current anti-podes are situated. Near each end of such an antenna, the amplitude of the current standing wave falls rapidly to zero at the outer extremities; consequently, the effective radiation from these parts of the antenna is minimal.

The antenna may also be used in the form of an inverted V. However, it should be remembered that for such a configuration to radiate at maximum efficiency, the included angle at the apex of the V should not be less than 120 degrees.[3] The use of 14 AWG enameled copper wire is recommended for the flat-top or V, although thinner gauges such as 16 or even 18 AWG can be used.

The Matching Section

This should be, preferably, of open-

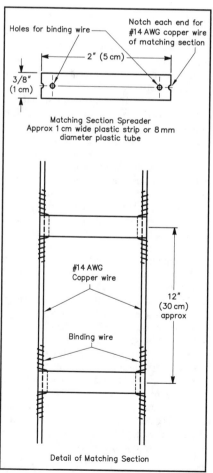

Fig. 10—Constructional details of the matching section. Also suitable for open-wire feeder construction.

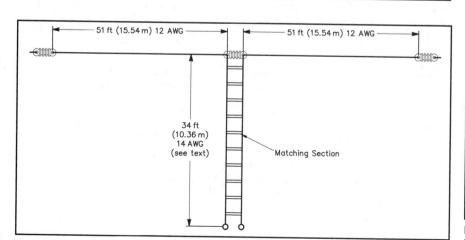

Fig. 9—Construction dimensions of the G5RV antenna and matching section.

wire feeder construction for minimum loss. Since this section always carries a standing wave of current (and voltage), its actual impedance is unimportant. A typical, and satisfactory, form of construction is shown in Fig. 10. The feeder spreaders may be made of any high-grade plastic strips or tubing; the clear plastic tubing sold for beer or wine syphoning is ideal.

If you decide to use 300-ohm ribbon type feeder for this section, it is strongly recommended that the type with "windows" be used. It has lower loss than a feeder with solid insulation throughout its length, and it possesses relative freedom from the detuning effect caused by rain or snow. If this feeder is used for the matching section, allowance must be made for its velocity factor (VF) in calculating the mechanical length required to resonate as a half-wave section *electrically* at 14.15 MHz. Since the VF of standard 300-ohm ribbon feeder is 0.82, the *mechanical* length should be 28 ft. However, if 300-ohm ribbon with windows is used, its VF will be almost that of open-wire feeder, say 0.90, so its mechanical length should be 30.6 ft.

This section should hang vertically from the center of the antenna for at least 20 ft or more if possible. It can then be bent and tied off to a suitable post with a length of nylon or terylene cord at an above-head height. Supported by a second post, its lower end is connected to the feeder.

The Feeder

The antenna can be fed by any convenient type of feeder provided always that a suitable type of matching network is used. In the original article describing the G5RV antenna, published in the *RSGB Bulletin* for November 1966, it was suggested that if a coaxial cable feeder was used, a balun might be employed to provide the necessary unbalanced-to-balanced transformation at the base of the matching section. This was because the antenna and its matching section constitute a *balanced* system, whereas a coaxial cable is an *unbalanced* type of feeder. However, later experiments and a better understanding of the theory of operation of the balun indicated that such a device was unsuitable because of the highly reactive load it would see at the base of the matching or make-up section on most HF bands.

If a balun is connected to a reactive load with an SWR of more than 2:1, its internal losses increase. The result is heating of the windings and saturation of its core, if one is used. In extreme cases with relatively high power operation, the heat generated in the device can cause it to burn out. The main reason for not employing a balun in the G5RV antenna, however, is that unlike a matching network, which employs a *tuned circuit*, the balun cannot compensate for the reactive load condition presented to it by

the antenna on most of the HF bands, whereas a suitable type of matching network can do this most effectively and efficiently.

Experiments were conducted to determine the importance, or otherwise, of unbalance effects caused by the direct connection of a coaxial feeder to the base of the matching section. There was a rather surprising result. The research showed that the HF currents measured at the junction of the inner conductor of the coaxial cable with one side of the (balanced) matching section, and at the junction of the outer coaxial conductor (the sheath) with the other side of this section, are virtually *identical* on all bands up to 28 MHz, where a slight, but inconsequential difference in these currents has been observed. There is, therefore, no need to provide an unbal-

---

> *The antenna can be fed by any convenient type of feeder provided always that a suitable type of matching network is used.*

---

anced-to-balanced device at this junction when using a coaxial feeder.

The use of an *unbalanced-to-unbalanced* type of matching network between the coaxial output of a modern transmitter (or transceiver) and the coaxial feeder is essential. This is because of the reactive condition presented at the station end of this feeder, which on all but the 14-MHz band, will have a fairly high to high SWR on it. The SWR, however, will result in insignificant losses on a good quality coaxial feeder of reasonable length; say, up to about 70 ft. Either 50- or 80-ohm coaxial cable can be used. Because it will have standing waves on it, the actual characteristic impedance of the cable is unimportant.

Another convenient feeder type that can be employed is 75-ohm twin lead. It exhibits a relatively high loss at frequencies above 7 MHz, however, especially when a high SWR is present. I recommend that not more than 50 to 60 ft of this type be used between the base of the matching section and the matching network. The 75-ohm twin lead available in the United Kingdom is of the *receiver* type; less lossy *transmitter* type is available in the United States.

By far the most efficient feeder is the

open wire type. A suitable length of such can be constructed in the same manner as that described for the open-wire matching section. If this form is employed, almost any length may be used from the center of the antenna to the matching network (balanced) output terminals. In this case, the matching section becomes an integral part of the feeder. A convenient length of open-wire feeder is 84 ft. It permits parallel tuning of the matching network circuit on all bands from 3.5 to 28 MHz, and with conveniently located coil taps in the matching network coils for each band, or where the alternative form of a matching network employing a three-gang 500 pF/section variable coupling capacitor is used, the optimum loading condition can be obtained for each band.[4] This is not a rigid feeder-length requirement, and almost any mechanically convenient length may be used. Since this type of feeder will always carry a standing wave, its characteristic impedance is unimportant. Sharp bends, if necessary, may be used without detriment to its efficiency. It is only when this type of feeder is correctly terminated by a resistive load equal to its characteristic impedance that such bends must be avoided.

Coaxial Cable HF Choke

Under certain conditions a current may flow on the outside of the coaxial *outer conductor*. This is because of inherent unbalanced-to-balanced effect caused by the direct connection of a coaxial feeder to the base of the (balanced) matching section, or to pickup of energy radiated by the antenna. It is an undesirable condition and may increase the chances of TVI [from fundamental overload, if the feeder is routed near a TV receiving antenna. — *Ed.*] This effect may be reduced or eliminated by winding the coaxial cable feeder into a coil of 8 to 10 turns about 6 inches in diameter immediately below the point of connection of the coaxial cable to the base of the matching section. The turns may be taped together or secured by nylon cord.

It is important that the junction of the coaxial cable to the matching section be made thoroughly waterproof by any of the accepted methods. Binding with several layers of plastic insulating tape or self-amalgamating tape and then applying two or three coats of polyurethane varnish, or totally enclosing the end of the coaxial cable and the connections to the base of the matching section in a sealant such as epoxy resin are a few methods used.

References

[1] Varney, L., "ATU or astu?," *Radio Communication,* August 1983.
[2] See Ref. 1.
[3] Varney, L., "HF Antennas in Theory and Practice— A Philosophical Approach," *Radio Communication,* Sept. 1981.
[4] See Ref. 1.

By William J. Lattin, W4JRW   From *QST*, April 1961

Multiband Antennas Using Loading Coils

Two-band operation can be obtained by using plain loading coils, with considerable constructional simplification as compared with the equivalent trap arrangement. This article discusses the principle, and gives dimensions for several 3.5-7 MHz combinations.

Many amateurs operate from locations at which it is impossible to put up a full-length doublet antenna for 80 meters. A doublet antenna can he shortened as much as desired by thc use of loading coils. The effect of loading coils is discussed very completely, with graphs and formulas, in *Bureau of Standards Circular C74, Radio Instruments and Measurements*, published in 1924 and reprinted in 1937. (Many an old-timer in radio will remember this as a standard reference book back in the '20s and '30s.)

It is shown that in addition to decreasing the natural frequency of an antenna, thc use of loading coils results in the fact that "the harmonic frequencies are no longer integral multiples of the fundamental as in the case of the simple antenna." In Fig 62, page 76 of the *Circular*, a graph shows how the next higher resonant frequency differs from the fundamental in one particular setup.

An antenna for 80 and 40 meters was made up according to this principle. A few trials with various values of loading inductance indicated experimentally that with 120-microhenry coils placed as shown in Fig 1, resonance occurred near the lower ends of both bands. With a small change in lengths, as shown in Fig 2, an antenna which resonated higher in both bands was obtained. Another small change in lengths resulted in the antenna shown in Fig 3, which is more satisfactory for phone operation. This antenna is 77 feet long, plus the lengths of the coils and insulators.

The coils were close-wound with No. 18 Nyclad wire on bakelite tubing $7/8$ inch in outside diameter, 14 inches long. A winding length of 12 inches was used. These coils measured approximately 120 µH. Some other coils were tried, 80 µH being the lowest value. Resonance in both bands was obtained but with longer lengths of wire. If the inductance Lof the coils is too low, the resonance at 40 meters may be too

high in frequency, although the 80-meter resonance can be gotten with longer lengths of wire on the ends. With various values of coils and lengths of wire, antennas can be made for 80 and 20, 80 and 15, 80 and 10, 40 and 20, and similar contributions.

As an antenna is made shorter it has sharper resonance. This may not be too much of a handicap for hams who operate over only 100 kHz or so in the 80-meter band, as many s.s.b. addicts do. The antenna of Fig 3 is actually just slightly longer than a regular doublet at 40 meters, up to the loading coils, and can be operated over the entire 40-meter band with a fairly low s.w.r. on the feeder. The advantage is two-band operation with an antenna 77 feet long without traps.

This antenna has been used on the air for several years and the reports have always been just about the same as those obtained with regular doublets. Obviously the loading coils should be made as low-loss as posible by using good insulation and as large wire as is practical. There are no capacitors to break down as in traps, and the 120-µH coils have been used with a kilowatt transmitter input with no difficulty.

We have not found any exact formulas to determine the relationship between the lengths of wire, loading coils, and the two frequencies. The antennas are very simple to adjust with a grid-dip meter coupled to a single-turn loop conected to the feed terminals, as quite small changes in the wire lengths result in appreciable changes in resonant frequencies.

This principle can be extended: that is, by using two sets of coils, operation on three frequencies is possible, on four frequencies with three sets of coils, and so on. However, these get very complicated to adjust, since the second set of loading coils changes operation of the first set somewhat, and the adjustment process gets rather tedious.

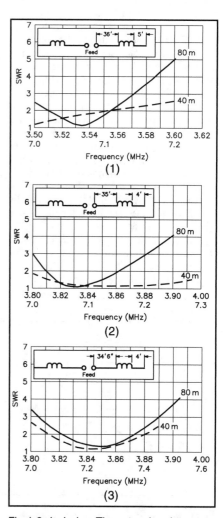

Fig 1-3, inclusive—Three two-band antenna configurations using 120-µh loading coils, showing effect of small variations in the lengths of the straight portions of the antenna. Dimensions and construction of the sides to the left of the feed terminals are identical with those shown to the right. Standing-wave ratio measurements made with RG-8/U cable (52 ohms) and Micromatch.

Easier Adjustment for the Two-Band Coil-Loaded Antenna

In an April 1961 *QST* article ("Multiband Antenna Using Loading Coils"),William J. Lattin, W4JRW, described how a horizontal dipole could be used on two adjacent HF bands merely by adding loading coils (Figures 1 through 3). The length of the antenna's inner section, adjusted for resonance at the higher frequency of the two bands, is just a bit longer than the usual dipole at that frequency. The element sections outside of the loading coils are trimmed for resonance at the lower of the two bands For 40/80-meter operation, he found that coils of about 100 to 120 µH work well. Lattin reported that depending on where in the two bands best SWR with 52-Ω coax is wanted, the inner section's length varied from 34 feet, 6 inches, to 36 feet, and the outer section's length varied from 4 to 5 feet.

For the same overall antenna length, it can be shown that the efficiency on the lower frequency band will be better than achieved with a 40-meter trap designed for the same outer section length, assuming that coils of equal Q are used. (The efficiency depends on coil Q.) Also, as Lattin writes, "There are no capacitors to break down as in traps, and the 120-µH coils have been used with a kilowatt transmitter input with no difficulty."

My space is limited, but I wanted to add

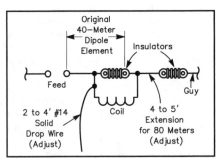

Fig 4—Solder the drop wire to one end of the coil, and the other end of the coil to the extension wire. Add the end insulator and attch the assembly to the 40-meter insulator and the guy line. Trim each to resonance in its band. (drawing not to scale)

80-meter capability to my present 40-meter half-wave dipole for those high-angle, short skip QSOs. From past experience with the arrangement, I knew it was a nuisance to adjust the inner section length by splicing and trimming. I didn't want to splice in the short inner section's increased length, so I tried the arrangement shown in Figure 4. This system required only the attachment of an assembly (consisting of the coil, drop wire, antenna extension and insulator) to my existing 40-meter antenna and guy line.

With this arrangement in place, I found adjustment by trimming to be much easier. Since antenna lengths will vary with the antenna's height above ground and surrounding objects, and with the inductance used, the dimensions given in Figure 4 include wide tolerances for adjustment. As Figures 1 through 3 indicate, the antenna's bandwidth on 80 meters is quite narrow. This makes tuning rather critical, both as to length and to equality of the extension lengths.—*Charles J. Michaels, W7XC, Phoenix, Arizona*

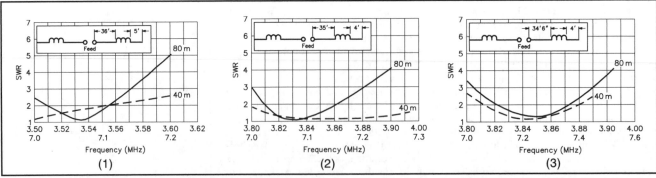

(1) (2) (3)

Figures 1 through 3 (l-r)—"Three two-band antenna configurations using 120-µH loading coils, showing the effect of small variations in the lengths of the straight portions of the antenna. Dimensions and construction of the side to the left of the feed terminals are identical with those shown to the right. Standing-wave measurements were made with RG-8/U cable and a Micromatch.... The coils were close-wound with No. 18 Nylclad wire on bakelite tubing ⅞ inch in outside diameter, 14 inches long. A winding length of 12 inches was used. These coils measured approximately 120 µH. Some other coils were tried, 80 µH being the lowest value. Resonance in both bands was again obtained, but with longer lengths of wire. If the inductance of the coils is too low, the resonance at 40 meters may be too high in frequency, although the 80-meter resonance can be gotten with longer lengths of wires on the ends. With various values of coils and lengths of wire, antennas can be made for 80 and 20, 80 and 15, 80 and 10, 40 and 20, and similar combinations.... The antenna of Figure 3 is actually just slightly longer than a regular doublet at 40 meters, up to the loading coils, and can be operated over the entire 40-meter band with a fairly low SWR on the feeder. The advantage is two-band operation with an antenna 77 feet long without traps." (from Lattin, April 1961 *QST*) Charlie Michaels simplifies the system's adjustment as shown in Figure 4.

By Chester L. Buchanan, W3DZZ · From *QST*, March 1955

The Multimatch Antenna System

Unique design providing essentially constant impedance over several bands

For a long time, hams have been searching for a single antenna that could be fed efficiently with a low-impedance transmission line on several bands. At last a simple but ingenious design by W3DZZ provides a solution. He has applied some well-known but neglected principles to both wire and parasitic-beam antennas.

Radio transmitters and receivers have enjoyed rapid development in flexibility to the point where changing bands is a matter of only spinning a dial or two and flipping a couple of switches. In contrast, the operation of a single antenna on several bands is usually done only at the expense of high standing waves on the feed line, because of the wide variation in antenna feed-point impedance from band to band.

Some work done by the author several years ago in connection with a dual-band parasitic array[1] has led to the development of a simple wire antenna covering five bands, from 80 to 10 meters. This antenna can be fed with a low-impedance transmission line without incurring excessive s.w.r. on any of these bands.

Basic Design

The fundamental principle of the system can be explained with the aid of Fig. 1.

In Fig. 1A, sections h_1 constitute a half-wave dipole for some frequency f_1. This dipole is terminated in lumped-constant trap circuits resonant at f_1. Additional wire sections, h_2, extend beyond the traps. If the system is excited at frequency f_1, the traps serve to isolate the dipole much as though insulators were inserted at these points.[2]

Low impedance at the center feed point of the antenna occurs not only at its fundamental resonance but also at any odd harmonic of the fundamental.

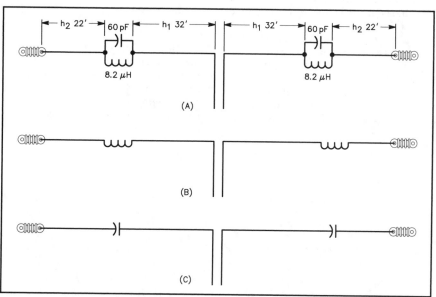

Fig 1—Sketch illustrating the three fundamental modes of the multimatch antenna.

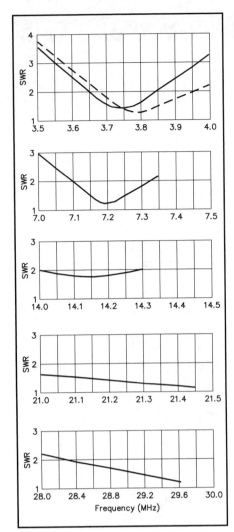

Fig 2–S.w.r. measurements made on the antenna of Fig 1A. The dashed lines show measurements made on a 122-foot dipole in the same location for comparison.

At frequencies much lower than f_1, the traps no longer isolate the dipole, but act simply as loading inductances in a second dipole whose electrical length is made up of h_1, h_2 and the inductive reactance of the traps, as in Fig. 1B.

At frequencies much higher than f_1, the traps again cease to isolate the sections, the traps now acting as series capacitances, as in Fig. 1C.

Another important consideration in this multiband system is that low impedance at the center feed point of the antenna occurs

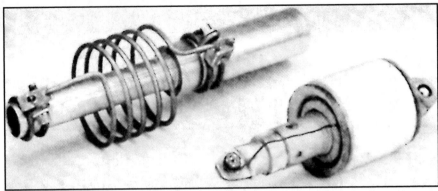

Lightweight weatherproof traps made by the author. To the left is the type inserted in beam elements, while the other one is suitable for wire antennas.

not only at its fundamental resonance but also at any odd harmonic of the fundamental.

By applying these principles, and by proper selection of the values of L and C in the traps, and choice of lengths for h_1 and h_2 it has been possible to arrive at a design where the system operates as follows:

1) Sections h_1 form a half-wave dipole resonant in the 40-meter band. The traps, resonant at the same frequency, isolate this dipole from the outer sections.

2) The inductive reactance of the traps is such that the entire system, including

This antenna can be fed with a low-impedance transmission line without incurring excessive s.w.r. on any of these bands.

sections h_2, resonates as a loaded half-wave dipole for the 80-meter band.

3) The capacitive reactance of the traps at higher frequencies is such that the entire system resonates as a $^3/_2$ wavelength antenna on 20, $^5/_2$ wavelength on 15, and $^7/_2$ wavelength on 10 meters.

The antenna is fed with 75-ohm Twin-

Lead, and Fig. 2 shows the results of s.w.r. measurements made across each band. Proper dimensions are given in Fig. 1A.

Trap Construction

The values of C and L used in the traps are quite critical. The capacitance should first be adjusted accurately to 60 μμf., then the inductance should be trimmed until the trap resonates at 7200 kHz. This should be done before the traps are inserted in the antenna. The inductance will be approximately 8.2 μH. The traps made by the author are 6 inches long and weigh only 6 ounces and the Q is well over 100. They will withstand the voltage developed by a 1-kW transmitter. Samples are shown in the photograph. The wire-antenna capacitor is made up of concentric lengths of 1-inch and $^3/_4$-inch aluminum tubing separated by polystyrene tubing with $^1/_8$-inch walls molded around the inner conductor. The polystyrene is also flowed into a series of holes in one end of the outer conductor so that the strain of the antenna will not pull the assembly apart. The inductor is wound with No. 14 wire and is concentric with the capacitor. The inductor is weatherproofed by molding it in insulating material. Other construction might be used, of course. As an example, a conventional inductor and capacitor could be enclosed in a plastic box, suspended across an insulator. This would, however, add to the weight.

Notes

[1]Buchanan, "Duo-Band Ham Antenna," *Radio & Television News*, December, 1950.
[2]Morgan, "A Multifrequency Tuned Antenna System," *Electronics*, August, 1940.

By J. R. Mathison, WB9OQM   From *QST*, February 1977

Inexpensive Traps for Wire Antennas

For the amateur who does not have everything but who enjoys making equipment, this easy approach to constructing a trap antenna pays off.

Not every amateur is blessed with ample room to install an antenna for the lower bands. For the higher bands there is also the matter of cost where the luxury of a tower and a beam outstrip the family budget. An alternative is to construct a wire antenna which contains traps an industrious amateur can make with a minimum of cost.

The only materials required for each trap are a small piece of $1/4$-inch-thick Plexiglas, a length of bare No. 12 copper wire, and a couple of feet of RG-8/U or RG-11/U coaxial cable. The Plexiglas serves as a strain insulator and coil form. Ceramic insulators are not needed for the traps. The coaxial cable functions as a high-voltage capacitor, eliminating the need for obtaining expensive commercially made capacitors for the trap assemblies.

Getting Started

To find a suitable antenna using the trap design, it is suggested that the amateur refer to the five-band dipole described in the 1977 edition of *The Radio Amateur's Handbook*. It will be noted that this particular antenna uses two 10-µH trap coils consisting of 15 turns of No. 12 wire, $2^1/2$ inches in diameter and with 6 turns per inch. Across each coil is a 50-pF capacitor.

To begin making the traps, a piece of Plexiglas is cut to form two rectangles of equal dimensions, 2×3 inches. The photograph of the sample trap in Fig. 1 shows that holes have been drilled in the Plexiglas to accommodate the coil wire. These holes are $1/4$-inch from the lengthwise edges, and they should be slightly larger in diameter than that of the wire to be used for the inductor. Also, to make winding easier, stagger the set of holes on one edge from those

Construction of the trap is shown with the coaxial-line capacitor dressed from the shield end of the coil.

on the opposite edge of the Plexiglas by half the turn spacing.

Prepare the wire by straightening any kinks or bends. The wire is then wrapped in one layer tightly around a cylindrical form. The form should be of a size that will allow the coil of wire, when released, to be approximately $1/16$-inch smaller in diameter than the required coil. The coil of wire thus formed is then threaded into the drilled Plexiglas rectangle in the same fashion as a wire spiral is threaded into a spiral notebook.

The Coaxial Capacitor and Coil

Drill two more holes in the centerline of the Plexiglas at the position of each end of the coil wire. One hole is for the antenna wire. The other hole is for a wire soldered to each end of the coil and anchored through the Plexiglas. These wires serve as an adjustable tap for each end turn of the coil and as a terminal strip for the antenna wire and coaxial capacitor.

To determine the capacitance per foot of the coaxial cable to be used for the trap capacitor, a cable manufacturer's catalog

should be consulted.[1]

The capacitance required for the antenna trap shown in *The Radio Amateur's Handbook* is 50 pF. To compute the length needed for the coaxial capacitor, divide this value by the nominal capacitance per foot. For RG-8/U in this example the required length would be approximately $20^5/16$ inches, and $29^1/8$ inches for RG-11/U. Be sure to allow a couple of extra inches for lead, as it is the braid length *remaining on the cable after preparation* which will determine the effective capacitance.

Final Touches

Trim one end of the coax so that the inner insulation and center conductor can be passed through the center of the coil and soldered on a radial tap. The braid is then soldered to the other radial tap. Use heat sinks to avoid melting the plastic. Cut off the other end of the coaxial cable a half inch longer than required. Use a tubing cutter to trim the outer insulation and braid from the remaining half inch. This will allow a longer insulating path in the air between the center conductor and braid than a square end cut. For other applications where air-wound coils are required and excessive heat is not a problem, this type of construction is suitable.

Resonance can be adjusted either by moving the tap wire or by pruning the coax capacitor. Install the antenna wire after adjusting the trap to the desired frequency. The coaxial-line capacitor is then taped to the antenna wire to support its weight and prevent straining the connections.

[1][Editor's Note: For RG-8/U the nominal capacitance per foot is 29.5 pF. For RG-11/U the nominal capacitance per foot is 20.6 pF.]

By Robert H. Johns, W3JIP

From *QST*, November 1983

Dual Frequency Antenna Traps

Although L-C antenna traps have been around for years, you've never seen any like these!

H ere is a new way to make antenna traps using only coils, without scarce and expensive high-voltage capacitors. An additional bonus is that these traps can be made to resonate simultaneously on two frequencies, greatly expanding their capabilities!

Cross-Linked Polyethylene (XLP) Insulation

The key to these new traps is a specially insulated wire that withstands several kilovolts. This wire is wound one layer on top of another, which produces some capacitance between the layers (Fig. 1). Enough capacitance in parallel with the coil inductance produces a resonant circuit, which can be used as an antenna trap. In transmitting service, this capacitor (formed by the insulation) will need to withstand high voltage and high currents without much dielectric loss. XLP, a recently developed type of insulation, has the excellent high-voltage and low-dissipation properties of polyethylene. It is also tough and hard from additional polymerization (or cross linking) of the molecules.

All of the traps described in this article are made from wire covered with XLP insulation. It is normally used for telephone switchboard service and costs about 10 cents a foot for no. 14 stranded, type-SIS wire.[1,2]

How They Work

In Fig. 1, notice that the two coils are

Dual-frequency resonance can be used to good advantage in a trap antenna.

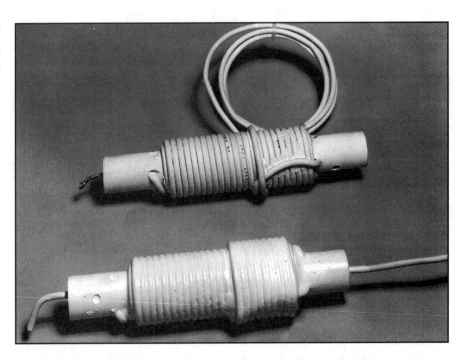

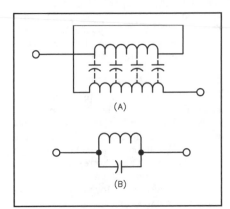

Fig 1–Illustration of distributed capacitance between coil turns (A). Electrical equivalent circuit of A (B).

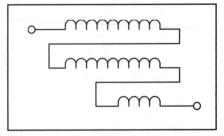

Fig 2–A dual-frequency antenna trap. The lower resonant frequency is determined by the entire three-layer coil and distributed capacitance. The higher-frequency resonance is determined by the smallest coil and associated capacitance. The two circuits interact, but can be adjusted individually by varying the number of coil turns.

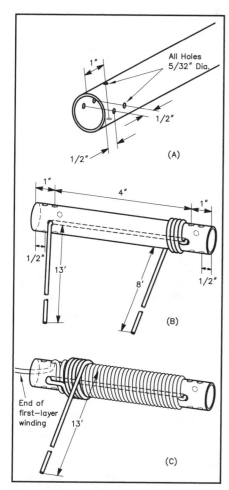

Fig 3—Construction details for a 40/15-meter trap. All text instructions follow this orientation of the coil form. Holes should be drilled as shown in A. First-layer coil winding is shown in B, and second-layer details in C.

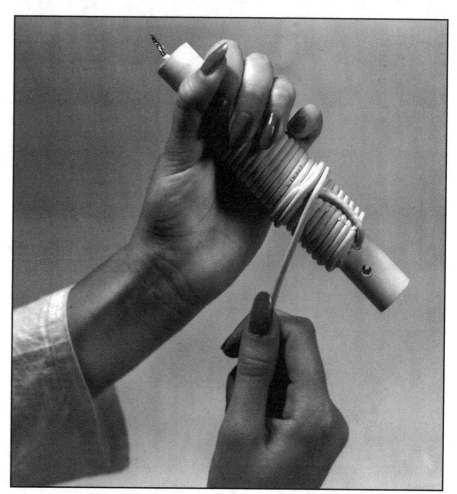

Fig 4—The third-layer coil winding technique for a 40/15-meter trap. This coil is wound on top of the wire lying over the second-layer coil.

wound in bifilar fashion. The more turns added to the two coils, the lower the resulting frequency, since inductance and capacitance are increasing simultaneously. This tuned circuit is similar to the filters described by Doty, and is an extension of the coaxial cable traps of O'Neil and Johns.[3,4,5]

> *A second resonant frequency is achieved by winding a small third layer on top of the first two.*

A second resonant frequency is achieved by winding a small third layer on top of the first two (Fig. 2). Another tuned circuit is formed by the capacitance between the outermost coil and the two others, together with the inductance of the small coil. This higher-frequency resonance appears only if the third coil is less than half the length of the larger coils. If too many turns are added, this resonance gets broader and shallower, finally disappearing when the coil becomes too large.

Dual-frequency resonance can be used to good advantage in a trap antenna. The common trap dipole for 80 and 40 meters will also work on 15 meters if the traps are tuned to 21 MHz. The half-wave section on 40 meters will function as three half-wave elements on 15 meters, and can also be fed at the center.

Construction Principles

The basic trap is a 40/15-meter version, and is made from $1/2$-inch PVC pipe. Cut the PVC pipe to 6-inch lengths. One length is needed for each trap. Drill holes in the pipe, as shown in Fig. 3A. The form is now ready to accept the winding.

Start winding the trap by passing a 21-foot length of wire through the center of the form. Pass the ends out through a set of holes spaced 4 inches apart. One end of the wire should protrude approximately 8 feet from the form, the other approximately 13 feet. With the 8-foot wire on your right, wind it toward the left edge of the pipe for a total of $25^{1}/_4$ turns (Fig. 3B). Feed the end of the wire through the appropriate hole and out through the center of the trap to form a pigtail. Now coil the other wire on top of the first layer by winding in the opposite direction of rotation, laying each turn in the spaces between first-layer turns (Fig. 3C). Skip a space at the beginning of the second layer, for it has only 23 turns. This coil should finish at the right side of the trap. Feed the remainder of the wire completely through the trap form by passing it through two opposing holes.

Take the end of the second-layer wire (there should be about 2 feet left) and lay it perpendicular to the second-layer turns. Count 10 spaces from the right end of the outer coil and make a 90° bend in the wire at this point (Fig. 4). Lay the wire into this space and wind $9^{3}/_4$ turns over the second layer, making sure the coil is following the same direction of rotation as the second layer. The third layer is actually wound over the wire that lies perpendicular to the coils. This outermost coil should end at the right side of the form. Pass the remaining wire through the appropriate hole and out through the center of the trap. You should now have a completed trap with three layers of windings and a pigtail of wire protruding from each end.

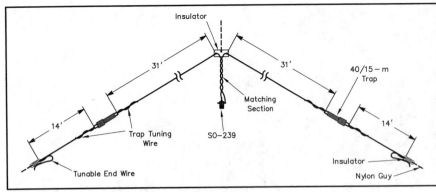

Fig 5–Dimensions of an 80/40/15 and 10-meter antenna using one pair of 40/15-meter traps. The center insulator can be made from 1/2-inch PVC pipe, and the SO-239 housing from 1-inch PVC slip caps. The 14-foot dimension includes the length of wire folded back for tuning. See text for details.

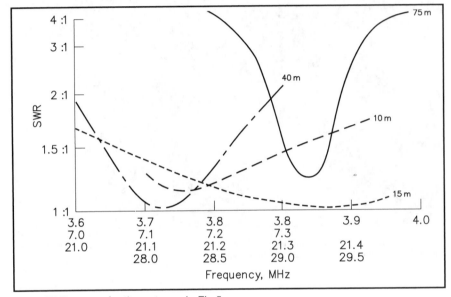

Fig 6–SWR curves for the antenna in Fig 5.

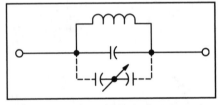

Fig 7–The trap-tuning element is a 3-foot length of no. 14, XLP-insulated wire. It is fed through the trap and loosely wrapped around the antenna wire on either side. At A, the capacitance is maximum and the resonant frequency will be higher than in A. A 3-foot wire can tune the antenna over 500 kHz on 40 meters. See text.

Fig 8–Electrical equivalent of a trap and tuning wire. The tuning wire forms a differential capacitor (dashed lines), which changes in series value as the wire is slid through the trap.

A Four-Band Antenna

Fig. 5 shows the dimensions of a trap dipole using one pair of the 40/15-meter traps described above. It is resonant on 80, 40, 15 and 10 meters.

The 10-meter resonance was a pleasant surprise, and exists because the entire antenna is resonant as five half-wavelengths, capacitively loaded by the traps. The $1^1/_2$-λ and $2^1/_2$-λ configurations on 15 and 10 meters, respectively, are not a good match to 50-Ω coaxial cable; a short matching section is used. This is made from the no. 14 SIS wire used in the traps. Two 6-foot pieces of wire lightly twisted together makes a $1/_4$-λ, 10-meter transformer (about 130-Ω) that also provides a good match on 15 meters. The matching section is so short that it doesn't affect the 40- and 80-meter bands significantly. The SWR curves of this antenna are shown in Fig. 6.

Tuning

Here is a tuning trick that can be used with any trap antenna, not just the ones shown here. An insulated wire is passed through or around a trap and capacitively coupled to the antenna wire on either side of the trap (Fig. 7) This effectively places a differential pair of capacitors in parallel with the trap capacitor. The equivalent circuit is shown in Fig. 8. When the tuning wire protrudes equal amounts from both sides of the trap, additional capacitance is at a maximum. In this case, the trap (and antenna) frequency will be brought down to the lowest possible value. Sliding the tuning wire to either side causes the differential pair to decrease in series value, thereby raising the trap (and antenna) resonant frequency. The tuning wire can therefore be slid just far enough to bring the antenna up to the desired resonant frequency.

When the 40/15-meter traps are tuned in this manner, 40 meters is affected the most, with 15 meters affected to a lesser degree.

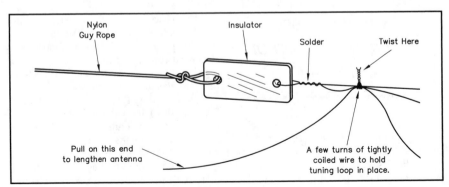

Fig 9–Method for tuning 80 and 10 meters on the four-band antenna in Fig 5. Resonant frequency is lowered by pulling out the free end of the loop.

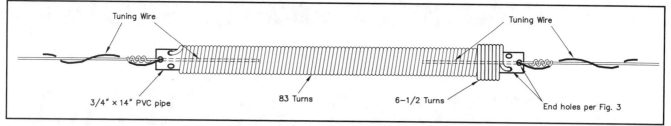

Fig 10–Dimensions of the 10/15-meter trap and tuning wires. The trap is wound in the same manner as the 40/15-meter version described in the text. Since this is only a two-layer coil, the end of the first layer is passed completely through the form and laid back over the windings. The second layer is then wound over this wire.

When constructing traps, the number of coil turns may need to be adjusted for your particular wire and layout.

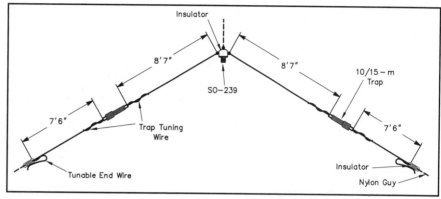

Fig 11–A 40/15- and 10-meter antenna using the traps shown in Fig 10. Because of the loading coils incorporated in the traps, this antenna is approximately half the size of a standard 40-meter dipole.

This works out well, since 15-meter resonance is usually quite broad and should not need much adjustment. Although traps should normally be constructed to resonate in the middle of the band, they should resonate at the top of the band when tuning wires are to be used. They will bring resonance back down into the center of the band.

To lower the resonant frequency on 10 and 80 meters, another simple tuning scheme is used. At the ends of the antenna, some of the wire is folded back along itself and then out through a knot at the end (Fig. 9). To lower the resonant frequency on 10 and 80 meters, some of this wire is pulled out and attached to the nylon guy rope. If the antenna is set up as an inverted V, this adjustment can be made without loosening the guys.

Other Frequency Combinations

Fig. 10 shows a 10- and 15-meter trap combined with a loading coil to shorten overall antenna length. The long coil is a single layer with a self-resonant frequency of approximately 29 MHz; it also serves as the loading coil. The smaller, second layer coil forms a 21-MHz trap. The outer portion of the antenna beyond the traps can be cut for any band lower in frequency than 15 meters (except 20 meters). Fig. 11 gives dimensions

Once a trap is connected into an antenna, all measurements must be done from the feed point.

for a 40/15/10-meter antenna. This antenna can also be tuned with insulated wires running through the traps. Two wires are used, one into each end of the trap (Fig. 10). The 15-meter tuning is quite critical. A $^1/_2$-inch shift in the tuning wire position moves the resonant frequency across the entire band!

Last-Minute Hints

When constructing traps, the number of coil turns may need to be adjusted for your particular wire and layout. Typical variations in the insulation thickness are enough to change the number of turns required. A GDO (grid-dip oscillator) is essential when

building traps, but don't try to measure the frequency of a trap with any wire connected to it or you will get an erroneous reading. Also, a trap cannot be dipped when it is in an antenna, for too many other resonances will appear. Once a trap is connected into an antenna, all measurements must be done from the feed point. When your traps are completed, wrapping them with electrical tape or dipping them in liquid silicone rubber to secure the turns in place is a good idea. This covering will also help protect the XLP insulation from the damaging effects of ultraviolet radiation.

Many features of the traps and antennas described here have been patented or are patent pending. Amateurs are welcome to build these for their own use, but manufacturers are cautioned that all patent rights will be strictly enforced.

Notes

[1]m = ft × 0.3048; mm = in. × 25.4.
[2]This wire is available from the local wire distributors and the Barker and Williamson Co, 10 Canal St, Bristol, PA 19007.
[3]A. Doty and A. Macnee, "Introducing the INCONS," QST, Feb 1979, pp 11-14.
[4]G. O'Neil, "Trapping the Mysteries of Trapped Antennas," Ham Radio, Oct 1981, pp 10-16.
[5]R. Johns, "Coaxial Cable Antenna Traps," QST, May 1981, pp 15-17.

By Al Buxton, W8NX   From *QST*, July 1996

An Improved Multiband Trap Dipole Antenna

You need this—traps with lower loss, higher Q, increased power-handling capability and four-band coverage!

This improved multiband trap dipole introduces a new trap design and a change in trap location. The antenna features *double-coaxial-cable-wound traps* having lower reactance and a higher quality factor (Q) than earlier coax-cable traps. Because trap loss resistance is determined by trap reactance divided by Q, these enhancements provide a substantial reduction in such losses. Of as much significance, the new traps have a *fourfold increase in power-handling capability* over that of other coax cable traps (more on that later). Weatherproof performance and the ease of construction of coax-cable traps is retained.

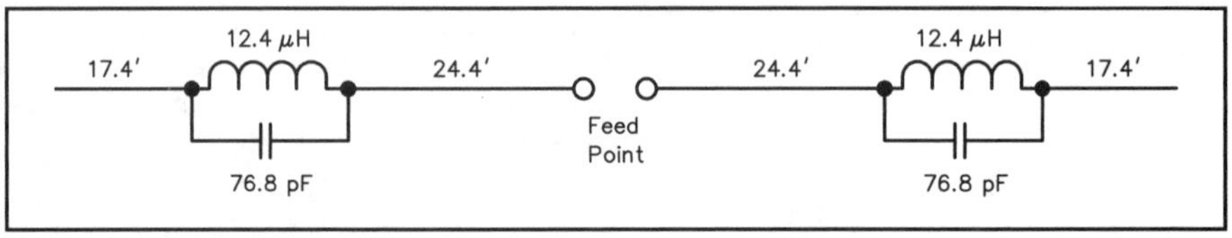

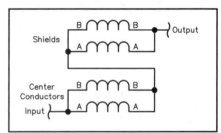

Figure 1—An improved 80, 40, 17 and 10-meter trap dipole.

(Figure 1 labels: 17.4', 12.4 µH, 24.4', Feed Point, 24.4', 12.4 µH, 17.4', 76.8 pF, 76.8 pF)

Figure 2—Close-up of a completed trap.

Describing how to make the traps is more difficult than making them!

Figure 3—Double-coax trap schematic.

(Figure 3 labels: Shields, B, B, Output, A, A, Center Conductors, B, B, Input, A, A)

even for those with few manual skills and tools. Figure 2 is a photograph of a trap; Figure 3 is the trap schematic. Double-coax-wound traps have *two parallel windings of coax cable* (A and B) rather than a single-cable winding. The two shield windings and center conductors are each connected in parallel. The series loss resistance of the trap—dominated by skin effect—is essentially halved by the parallel connection. The surface area for convection and heat conduction away from the trap is essentially doubled. Power losses in the polyethylene dielectric of the trap are almost negligible because of the relatively low frequencies involved.

The SWR performance of the antenna at my location, using a 50 to 75-Ω transformer and a 75-Ω feed line, is shown in Figure 4. The antenna is installed as an inverted **V** with a 40-foot apex height. Notice the very good performance on 40

This dipole (see Figure 1) operates on 80, 40, 17, and 10 meters; four of our more popular bands. The dipole is made of #14 stranded copper wire radiating elements, a 1:1 balun, a pair of insulators and a pair of the new traps. Notice the change in *trap location*. The traps are at the ends of 24.4-foot elements rather than 32-foot elements as in a conventional 80 and 40-meter trap dipole. Trap resonance on the four bands is nonexistent. Because there is no open-switch divorcement action by the traps, the full length of the antenna radiates on the four bands. Avoiding trap resonance lowers the possibility of trap-voltage break-

down. True resonant-current feed is obtained on the four bands, making the antenna compatible with either 50 or 75-Ω coaxial-cable feed lines. Fundamental-frequency operation is provided on 80 and 40 meters, and long-wire, odd-harmonic operation on 17 and 10 meters.

This antenna is a rewarding and inexpensive project for those of you who like to homebrew your own equipment. I designed the antenna with a little cut and try using my new *Trap and Stub Dipole Antennas for Radio Amateurs* computer software package.

Trap Construction

Trap construction is relatively easy,

meters, where the SWR is less than 2 across the entire band. The antenna favors the low end of 10 meters, where most of the activity seems to be concentrated. On 17 meters, the SWR is close to 3 across the band, requiring an antenna tuner to keep the rig happy. The 2:1 SWR bandwidth on 80 meters is 75 kHz, centered on 3.79 MHz. A good antenna tuner can extend the operating bandwidth on 80 meters well into the CW portion of the band or into the General class phone portion.

I measured the Q of these traps at two widely separated low frequencies and extrapolated the results to the higher operating frequencies. This two-frequency extrapolation method solves an otherwise impossible problem of directly measuring the Q of coaxial-cable traps. This approach separates the dielectric losses and the skin-effect ohmic losses of the trap; it assumes the skin-effect losses vary as the square root of frequency, and the dielectric losses vary as the first power of frequency. The results are shown in Figure 5, where the Q of an RG-59 double-coax trap is compared with the Q of a common RG-59 coax trap. The Q of a common RG-58 coax trap is also shown. The superiority of the double-coax configuration is clearly evident. The superiority of RG-59 over RG-58 is also demonstrated. All traps are tuned to a nominal 5.16 MHz and are of the optimum Q and length/diameter ratio. As you can see, the double-coax RG-59 configuration has a Q about 18% higher than the common singly wound coax configuration.

I calculated the trap losses for the bands covered using Roy (W7EL) Lewallen's *EZNEC* program. The results are shown in Table 1. The antenna was assumed to be horizontally mounted, 40 feet above real ground and using the legal power-output limit of 1500 W PEP. The stray capacitance

of the outer shields of the coax traps was simulated as a 1.5-foot length of #14 wire hanging down from the outboard end of the traps. The ratio of PEP to average power is assumed to be 2:1, corresponding to ideal two-tone single-sideband modulation. Notice the very high radiation efficiency and negligible trap dissipation on 40, 17 and 10 meters. On 80 meters, however, the antenna's low height and its shortened length reduce the antenna input resistance to 51.8 Ω, making trap loss significant. On 80 meters, the trap Q is 171. The combined loss from both traps equals 1.66 dB on this band, an insignificant amount. The input-resistance component chargeable to trap loss becomes significant to the extent of dissipating 24.9% of the input power in the traps as heat. At this power level, there is an average dissipation of 119.4 W in each trap. The voltage drop across the traps is 2984 V, below the 3400-V rating of the cable.

My operating experience indicates that common coaxial-cable traps can dissipate 35 W or more without failure. Therefore, I believe that the double-coax traps can dissipate 140 W or more without failure, indicating that the 119.4-W dissipation per trap at the 1500-W PEP level can be handled successfully. Because I lack the necessary

amplifier, I haven't confirmed such operation. So far, I've operated this antenna only at the 600-W PEP output level of my Yaesu FL-7000 amplifier. (During subsequent ARRL Lab tests, these traps successfully handled a two-tone 1500-W PEP signal for 10 minutes with no signs of stress.—*Ed.*) If any amateur has a theoretical basis for calculation of trap power-dissipation capability, please let me know about it.

Details of the trap construction and its connections to the antenna segments are shown in Figures 6 and 7. Before commencing construction, study the trap details in these figures as well as Figures 2 and 3. The trap resonant frequency is 5.16 MHz. Trap resonant-frequency tolerance is 50 kHz, permitting selection of a resonant frequency anywhere between 5.11 and 5.21 MHz. The band most sensitive to trap-frequency error is 80 meters, where a trap-frequency error of 50 kHz causes an antenna resonant-frequency error of 30 kHz of the same sign. Thus, if you have a General class ticket and want to be above 3.85 MHz on 80 meters, it's better to have the trap resonant frequency err on the high side rather than on the low side of 5.16 MHz. Indeed, General class ticket holders may want to shorten the trap windings by an eighth of a double-turn to set the 80-meter antenna frequency above 3.85 MHz. In that case, increase the trap frequency to about 5.35 MHz. The trap frequencies should be checked with a dip meter and an accurate frequency reference before installing the traps in the antenna. Errors and changes in the trap's resonant frequency have much smaller effects on the 40, 17, and 10-meter bands than on the 80-meter band.

Making the Traps

Describing how to make the traps is more difficult than making them! See Figures 6 and 7. Each trap consists of 6.88

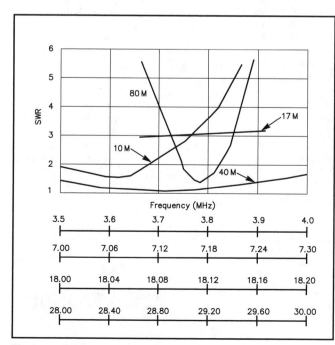

Figure 4—SWR plot of the improved trap dipole.

Band (meters)	Freq (MHz)	Radiation Efficiency (%)	Trap Loss (dB)	Trap Power Dissipation (W)
80	3.8	68.2	1.66	119.4
40	7.15	99.2	0.04	3.1
17	18.1	99.9	0.001	0.04
10	28.5	99.9	0.000	0.01

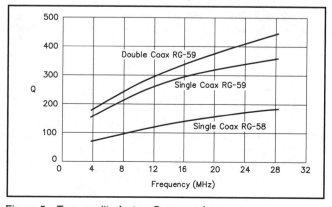

Figure 5—Trap quality factor, Q, versus frequency.

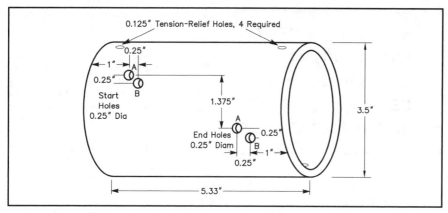

Figure 6—Hole positioning for the trap coil form.

Because there is no open-switch divorcement action by the traps, the full length of the antenna radiates on the four bands.

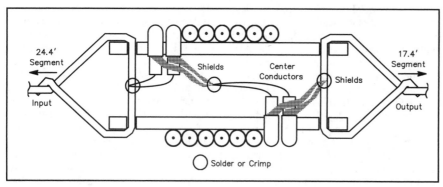

Figure 7—Cross-sectional view of a double-coax trap.

double turns of seven-foot lengths of RG-59 (Belden 8241) coaxial cable (A and B) wound on a PVC form.[1] Each form is a 5.33-inch length of 3.5-inch-OD schedule-40 PVC pipe (three-inch *ID* PVC pipe). Confirm the 3.5-inch OD of the PVC because trap frequency is sensitive to form diameter. Drill two 0.25-inch-diameter holes at the windings' start and end. See Figure 6. The center of the starting hole for winding A is one inch from the end of the form. Stagger the two starting holes relative to each other, using the hole for winding A as the reference location. The B winding starting hole is delayed 0.25 inch longitudinally and 0.25 inch along the circumference relative to the B winding starting hole. The 6.88 fractional number of double-coax turns is ¹/₈ of a turn less than a full seven turns. Therefore, the ending-turn holes are drilled ¹/₈ of the circumference (1.375 inches) shy of seven full turns. The 0.25-inch staggered relationship at the start of the windings is repeated at the end of the windings. Thus, the end winding hole for winding A lags 1.375 inches along the circumference behind the start winding hole for winding A.

Chamfer the hole edges with a sharp utility knife or rattail file to permit the cable to feed easily through the holes. The cable's 0.242-inch OD makes for a snug fit in the 0.25-inch-diameter holes.

The seven-foot cable lengths leave four inches at each winding end to make the trap

pigtails that extend inside the PVC form. Wind the A and B turns simultaneously. Use locking pliers inside the PVC form to firmly hold the two pigtails at the start of the windings. Clamp the far end of the windings in a vise to maintain firm and equal tension on the two cables as you wind them around the form. Avoid gaps between turns. After placing the turns, insert the cable ends through the end holes. While maintaining tension on the cables, use your fingers and needle-nose pliers to push and pull the cables into the inside of the form.

Using a sharp utility knife, strip away the last three inches of polyvinyl cover from the pigtails. *Avoid cutting the shield braid.* Separate the pigtail's shield braid from the dielectric and its inner conductor. Balloon the shield braid to an increased diameter by pushing it toward the end of the polyvinyl cover. Use a sharp pick to spread the braid and form a large hole close to the polyvinyl cover through which you can pull the cable's dielectric and center conductor. Again, be careful to avoid damaging the braid. The three-inch pigtail has now become two smaller pigtails—a shield pigtail and a center-conductor pigtail—each somewhat less than three inches long. Remove about two inches of the dielectric to expose the inner conductor. Convert all four ends of the cable windings into similar pigtail pairs.

The *center conductors at the start of*

the winding connect together, forming the trap *input* terminals for attachment to the 24.4-foot antenna wire. You may prefer to use crimp connectors rather than solder connections. Access to the inside of the traps—where the connections must be made—is somewhat awkward. The center conductors at the far end are fed back through the interior of the form, where they are connected to the shield braid at the start of the windings. The *far-end shield braids* connect together to form the *output* terminal of the trap for attachment to the 17.4-foot antenna segment. *It's important to avoid reversing the input and output terminals of the traps!* Reversal detunes the antenna a small amount by misplacing about 4 pF of stray capacitance of the trap shield braid. Build the second trap identical to the first.

Summary

Use a high-quality 1:1 balun for the antenna's center support and feed terminals. The 24.4-foot antenna segments attach to the balun's output terminals. Attach the trap-input pigtails to the far end of the 24.4-foot segments. Check the 24.4-foot segment lengths, measuring from the balun eyelets to the trap strain-relief terminals. The trap-output terminals connect to the near end of the 17.4-foot antenna segments. Check the 17.4-foot-segment lengths. Terminate the far ends of these segments in good-quality end insulators capable of withstanding the high RF voltage present at the antenna end points. The tension on the antenna may be loose, allowing considerable antenna sag. Install the antenna as high as possible—at least 35 feet above the ground at the center—and with the ends at least 15 feet high. Feed the antenna with either 50 or 75-Ω coaxial cable, the higher value being preferred to lower the feed-line SWR on 17 and 10 meters.

That done, go and have some fun!

Note

[1] Make sure the coax you use has a polyethylene (not foam) dielectric. A foam dielectric allows the center conductor to migrate.

CHAPTER THREE
LOOP ANTENNAS

By F.N. Van Zant, W2EGH   From *QST*, January 1973

160, 75, and 40 Meter Inverted Dipole Delta Loop

Presented here is a description of an efficient, full-size 160-meter antenna that fits in the same space as an inverted V, 75-meter dipole, with the capability of functioning as a full-wave triangle loop on 75 or 80 meters, as well as two full waves on 40 meters.

Now that the sunspot cycle is rapidly declining to a minimum, nocturnal activity on 10, 15, and 20 meters is curtailed proportionately and daytime activity on 10 meters is sporadic at best. Many amateurs are migrating back to the lower-frequency bands through necessity, if not by choice. Because of the sad state of 40 meters at night, with high-power broadcast stations every few kilohertz, the band that bears the burden of activity is 75 meters. Another amateur band that could shoulder some of the increased activity, but doesn't to any

Many amateurs are migrating back to the lower-frequency bands

great extent, is 160 meters. Several factors are responsible for this:

1) Most amateur equipment covers 80 through 10 meters only.

2) Antenna size and efficiency becomes a major stumbling block.

3) High atmospheric noise levels prevail in the tropical latitudes.

Factor 1 is easy to solve. One can modify existing equipment or resort to home construction. Factor 3 is not a major one to most U.S amateurs. Item 2 poses the major problem since many residential lots cannot accommodate a conventional full-size 160-meter antenna. Generally, a full-size 75-meter doublet or inverted V is possible. This article presents a practical solu-

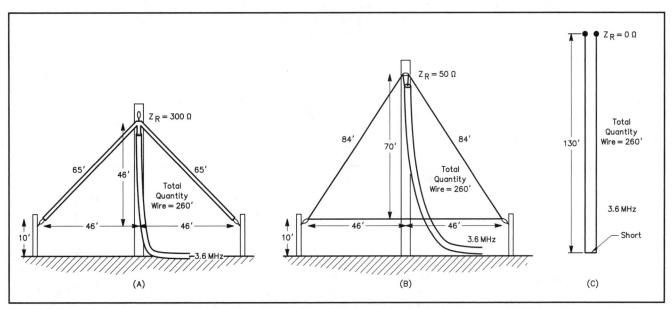

Fig 1—(A) A folded dipole for 75 meters, (B) a triangle with dimensions for a full-wave loop on 80 meters, (C) a quarter-wave section of transmission line.

tion for the operator who wants a full-size 160-meter antenna which occupies no more space than a 75-meter inverted V. No complicated counterpoises, antenna tuners, or open-wire feed lines are needed. As a bonus, efficient operation on 75/80 and 40 meters is possible. Only one 50- or 75-ohm feed line is used.

Antenna Development

For a number of years the author listened to 160 meters with envy while marveling at the uncrowded conditions existing in that relatively narrow allocation. In the prime evening hours of the winter months, it was not unusual to listen to uninterrupted QSOs between East Coast and Midwest stations that were running only 25 or 50 watts. These stations had one thing in common—a good antenna.

A full-size doublet on 1800 kHz is approximately 260 feet long and will not fit on a typical suburban lot (unless you build your home on an old railroad right-of-way!). An 80-meter folded dipole has a total of 260 feet of wire, yet fits in 130 feet of space. It can be erected on even less property if it is an inverted V as shown in Fig. 1 A. If the long second wire of the inverted-V folded dipole were dropped to horizontal, and the ends or top of the inverted V were adjusted to take up slack, a triangle would be formed as in Fig. 1B.

The height was adjusted to accommodate the same ground dimensions as the dipole of Fig. 1A; however, the example shown requires an 80-foot mast. As indicated in Table 1, the radiation resistance of the triangle has lowered to the 50- or 100-ohm region. If it were possible to keep raising the top of the triangle, pulling in the sides, the end result would be an 80-meter, half-wave, two-wire transmission line with the end opposite the feed point shorted, as shown in Fig. 1C. The feed point of the half-wave transmission line would have an impedance of zero ohms. This fact should be kept in mind, since it illustrates how the 80-meter impedance of the antenna may be adjusted by those who are fortunate enough to have a sufficiently high mast.

Figure 1B illustrates a full-wave triangle antenna for 80 meters. On 160 meters, however, it will be a half-wave triangle (total length of wire) having an extremely high impedance at the feed point. By simply opening the base wire at the center, the configuration becomes a 160-meter halfwave inverted-V dipole with the ends folded back toward the center mast. The antenna may be resonant at a slightly higher frequency than its calculated halfwave resonance. Operation on 160 or 80 meters can be ac-

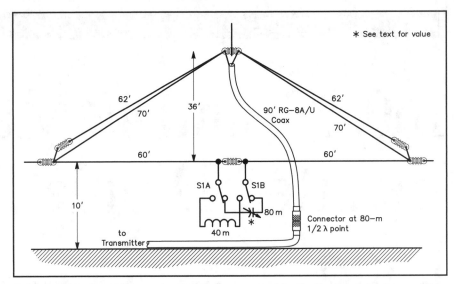

Fig 2–Suitable dimensions and switching arrangement for the three-band antenna. S1 is a dpdt knife switch. The switch is left in the open position for 160-meter operation.

complished by either opening or shorting the center point of the base of the triangle with a switch.

Since the author does not have an 80-foot mast the base ends of the triangle had to be pulled out to take up the slack. As shown in Fig. 2, the dimensions of the actual antenna approaches a more realistic value for many installations. The problem with lowering the height and pulling out the ends is that the impedance on 75 meters increases, mostly because of the antenna geometry, approaching that of a folded dipole again, With the dimensions shown in Fig. 2, the impedance on 80 meters is approximately 150 ohms. Two additional wires were added at the feed point to form a halfwave inverted V for 75 meters. The dipole elements illustrated in Fig. 2 resonate at 3900 kHz. A bonus feature of the triangle loop is it operates as a 2 wavelength resonance, with a 50-ohm impedance, at 7400 kHz. By adding a small loading coil to the center of the base wire, the resonance can be centered in the 40-meter band. To accommodate all three bands a dpdt knife switch can be mounted on the mast at the center of the horizontal wire. With the switch open, the 160-meter inverted V is active. With the switch in the 75-meter position, the full-wave triangle and the 75-meter dipole elements are active. With the

switch set for 40 meters, the loading coil is activated. The inductor for 40 meters should be about 20 turns of No. 12 or No. 14 bus wire, 2½ inches in diameter, 7 turns per inch, with taps and a clip lead for adjustment. Approximately 10 turns are needed to obtain resonance at the high end of 40 meters.

Table 1 presents measurements made with the 260-foot triangle antenna shown in Fig. 2. Since the antenna is fed at the top, measurements were made at the end of a 90-foot RG-8A/U feed line which rep-

Table 1

Switch Position	Calculated Freq. (kHz)	Measured Freq. (kHz)	Impedance	SWR 50-Ω Feed	SWR 70-Ω Feed
Open	1800	1825	50	1:1	1.4:1
80 Meters (without added dipoles)	3600	3700	100	2:1	1.4:1
80 Meters (with dipoles added)	3600 & 3900	3700 & 3900	50	1:1	1.4:1
40 Meters (with loading coil)		7250 (or any desired)	50	1:1	1.4:1

resents a half wavelength at 3.6 MHz and two half waves at 7.2 MHz. With this arrangement, the antenna feed-point impedance will be repeated at the far end of the feed line. On 160 meters, the line is flat and the resonant frequency is easy to determine. If the additional dipole elements are not desired on 75 meters, the antenna could be fed with 70-ohm coax.

Bottom Feed

A major advantage of the antenna configuration shown in Fig. 2 is the convenient access to the center of the triangle base for making adjustments. If it is desired to feed the triangle at the center of the base, some method would have to be devised to open and close the triangle at the top. An automatic passive switch at the top of the triangle can be fashioned from half-wave and quarter-wave transformers A 90-foot piece of coax cable (¼ wavelength long at 160 meters) shorted at the far end will present a very high impedance between the two elements of the antenna at the apex. The triangle then will look like a dipole on 160 meters. The same length of coax at 80 meters is a one-half wavelength shorted transmission line with the short being reflected to the apex of the triangle. On 80 meters the triangle will look like a continuous loop. Two half waves shorted provides a similar function for the 40-meter triangle. The same switch arrangement can be retained at the center of the triangle base

---

A full-size doublet on 1800 kHz is approximately 260 feet long and will not fit on a typical suburban lot

---

By simply opening the base wire at the center, the configuration becomes a 160-meter halfwave inverted-V dipole with the ends folded back

---

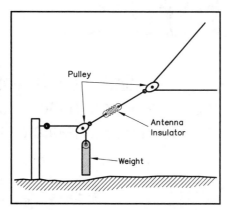

Fig 3—Pulley and counterweight arrangement to allow easy adjustment of the triangle area.

where the feed line attaches for switching in either the capacitor or inductor.

Beam Arrangements

The possibility of erecting two triangle antennas, appropriately spaced and parallel to each other has been considered. By tuning one of the triangles higher (or lower) than the band in use, a director (or reflector) is formed, thereby providing a beam pattern.

Construction

The antenna builder should consider a rope and pulley arrangement both at the top and at the ends of the triangle in order to adjust the height-to-base ratio to make impedance adjustments. The counterweight arrangement shown in Fig. 3 will keep tension on the base and top wires.

Soft drawn solid enameled or stranded No. 12 to 16 gauge copper wire is recommended in preference to Copperweld, which is quite stiff. The author's installation uses 16-gauge stranded, teflon-insulated wire purchased from a surplus dealer. A mast is the preferred center support; however, a tall tree will suffice. A bow, arrow, and light monofilament fishline can be used to shoot a pull line over a high limb.

Operation

Results with the antenna have been most gratifying. CW contacts as far as 400 miles were made using a 5-watt, 160-meter VFO for a transmitter. A-m contacts using a 5-watt transmitter were made as far as 300 miles in the 1825- to 1850-kHz segment. Performance on 75 meters has been consistently equal to that with any dipole antenna previously used over a 20-year period. Performance on 40 meters, at night, was a pleasant surprise. The 2-wavelength triangular loop seems to exhibit gain over a dipole, with low-angle radiation in the broadside direction. A number of QSOs have been made with W6 and W7 stations from New Jersey. For the suburban antenna experimenter and "top-band" enthusiasts, the three-band inverted dipole/delta loop seems a natural step for solving the problem of space and efficiency on the lower-frequency bands.

By Richard O. Gray, W9JJV

From *QST*, March 1983

The Two Band Delta Loop Antenna

An interesting antenna, indeed! But you may be more interested in the electronic switching for a remote impedance-matching network.

This antenna provides excellent performance on the 40- and 80-meter bands. It requires a limited amount of space, and includes a pretuned, remotely located antenna-matching network. The design illustrates the use of resonant circuits instead of switches to select the proper matching circuit for each band. You can build a similar antenna, with variations in loop shape or with different impedance matching circuits, if you follow the basic philosophy that I did.

Fig. 1 shows the arrangement I used in developing the matching networks. The open-wire feeder was connected to an 18-inch length of RG-8/U cable at the outside wall of my house.[1] This short length of coaxial cable served as a feedthrough line, so the experimental work could be carried out in the comfort of my home.

Only two pieces of test equipment—an rf voltmeter and an SWR indicator — were used to adjust the matching network. An exciter with a 50-ohm output impedance was adjusted to produce just enough power for a full-scale reading on the SWR meter. The rf voltmeter was connected between the inner conductor of the short coaxial cable and ground. The rf voltage at this point can be 1000 or greater when operating at full power, so be careful! Alternatively, a field-strength meter can be loosely coupled at this point. Short pieces of coaxial cable connect the transmitter to the SWR meter, and the meter to the matching network.

The total length of wire in the triangular loop is 140 feet, which is close to being a full wavelength on 40 meters and a half wavelength on 80 meters. I am sure the loop is affected by the close proximity of the vertical portion to the steel tower leg.

The apex of my loop is mounted by means of an insulator tied to a nylon rope through a pulley at the top of my tower. The rope is a continuous loop from the ground through the pulley and back to the ground. Should the wires break, or I have to lower the antenna for any reason, I won't have to climb the tower.

My matching network was determined empirically by trying parallel and series

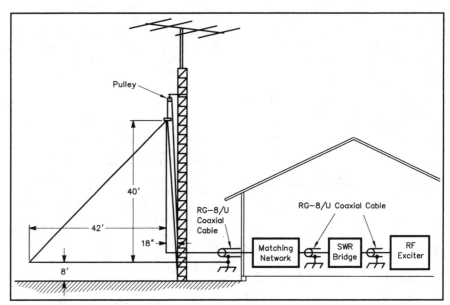

Fig 1—Construction details for the two-band Delta Loop. Notice that the matching network was located inside the house during the design stage. It was moved outside for the final arrangement.

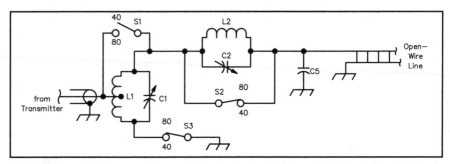

Fig 2—Schematic diagram of the matching network. Switches S1, S2 and S3 are shown in the 40-meter position.

tuned circuits, pi networks and various other circuits. You may find it easier to discover the best network for your particular case by using a commercial antenna-matching network that has most of the common combinations included. When a proper impedance match is obtained, you can build a pretuned network using the appropriate components and values.

All of the parts used in the experimental work came from my junk box. Inductors are wound on various tubing sizes that I had available. All of the wire for the loop, open-wire line and coils is no. 16 copper conductor.

To adjust the experimental matching network, I tuned the capacitors and changed the inductor tap points while watching the SWR meter and the rf voltmeter. Look for a decrease in SWR and an increase in the rf voltage. Patience and practice will lead to a low SWR and a fairly high rf voltage at your favorite operating frequency.

The End Result

In the final configuration, I moved the matching network to the outside wall of my house. Fig. 2 shows the final matching-network eireuit. Since most of the components are above ground potential, I used a 12- × 12-inch piece of $^1/_4$-inch-thick Lucite as a base. The enclosure is made of 0.018-inch-thick aluminum gutter flashing, available from most hardware stores. I no longer needed the 18-inch piece of RG-8/U to feed through the house wall, but it is still a part of the circuit capacitance. I connected one end of the center conductor to the network output, and the shield braid to ground. The other end of the cable is wrapped with electrical tape to prevent shorting to any other part of the circuit. This is C5 in Fig. 2.

Next, I simplified the operation by eliminating the three switches. I did this by using tuned circuits to switch the signal paths automatically. I determined that the impedance of Ll, and Cl is negligible on 80 meters when compared to that of L2, and C2. S1 can simply be eliminated. The S2 mode must present a high impedance on 80 meters and be near zero on 40 meters, so I replaced the switch and its function with a series-resonant circuit tuned to 40 meters. This circuit presents a capacitive reactance on 80 meters and reduces the required value of C2. The S3 mode should provide a low impedance on 40 meters and a high impedance on 80 meters. S3 can also be replaced by a series-resonant circuit that is tuned to 40 meters. The refined circuit is shown in Fig. 3. Fig. 4 shows the internal layout with the aluminum cover removed from the box.

Circuit Analysis

On 40 meters, the resonant impedance of L4 and C4 is about 3 ohms, assuming a conservative Q value of 20. This is considerably less than 50 ohms, and so meets my requirements. On 80 meters, the impedance is about 70 ohms, which is not considerably greater than 50 ohms. This impedance is a capacitive reactance, however, and can be tuned out by the adjustment of C2.

The resonant impedance of L3 and C3 on 40 meters is approximately 30 ohms, again assuming a Q of 20. This is much lower than the anticipated impedance of the antenna system. On 80 meters, the impedance is equivalent to a 60-pF capacitor, which is in parallel with C2. The capacitance of C2 will have to be reduced accordingly.

Final Adjustment

C1 and C3 should be adjusted for the lowest attainable SWR on 40 meters. Next, adjust C2 for a low SWR on 80 meters. Any

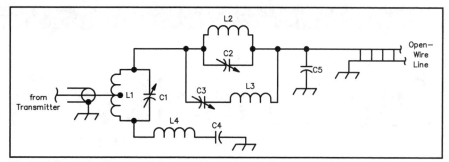

Fig 3—Schematic diagram of the matching network after the switches have been replaced with series-resonant circuits.

C1–10 to 80 pF, with 0.035-inch spacing between planes.
C2–21 to 100 pF, with 0.080-inch spacing.
C3–8 to 50 pF, with 0.080-inch spacing.
C4–400 pF, 1000 V.
L1–18 t, center tapped and tapped up two turns from bottom. The coil is wound on a $1^3/_4$-inch diameter form, and is 2 inches long (approx. 6.3 µH).
L2–19 t, tapped up five turns from input

end. The coil is wound on a $1^3/_4$-inch diameter form, and is 2 inches long (approx. 5.5 µH).
L3–20 t on a $1^7/_8$ inch diameter form, $1^5/_8$ inches long (approx. 12.0 µH).
L4–7 t on a $1^1/_{16}$-inch diameter form, $^5/_8$-inch long (approx. 1.35 µH).
All coils were wound with no. 16 copper wire.

Fig 4—Construction details of the automatically switched matching metwork for the two-band Delta-Loop antenna.

change in the setting of C3 on 40 meters will affect the setting of C2 on 80 meters. As a final step, switch between the 80- and 40-meter bands to be sure the performance is satisfactory on both.

The SWR curves for my system are shown in Fig. 5. If the frequency of lowest SWR is chosen properly, you should have an SWR of 2.5 or less across the entire 40-meter band, and for a 300 kHz segment of the 80-meter band.

Conclusions

My station is equipped with a 50-foot tower and is a triband Yagi antenna. The tower is grounded and is tuned for 40- and 80-meter operation by means of a gamma match. Tuning and switching are done remotely from the operating position. I used this as a vertical antenna for comparison with my Delta Loop.

The most noticeable difference between the antennas is in reception on 80 meters. The atmospheric and man-made local noise can be as high as S5 or more when using the vertical antenna. At times, I know there is a signal

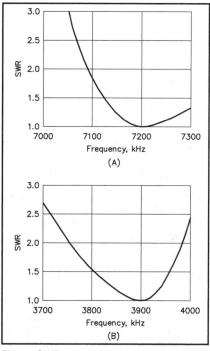

Fig 5—SWR curves for the W9JJV two-band Delta-Loop antenna. The 40-meter curve is shown at A, and the 80-meter curve is shown at B.

present, but I can't copy it through the noise. Upon changing to the Delta Loop, the station is "solid copy." This is what I was looking for.

I expected the loop to exhibit directional characteristics, with the maximum propagation being perpendicular to the plane of the loop. This was not borne out in the limited number of tests I conducted. The directivity might be affected by the tower, which is only 18 inches from the vertical side of the loop. It could also be affected by the aluminum siding on my house (10 feet away) or by the forest on the other side of the tower.

The receive sensitivity is about the same as with the vertical antenna. The Delta Loop may be $^1/_2$ S unit better. On transmit, the loop results in better reports by an average of 1 S unit.

By Dave Fischer, WØMHS   From *QST*, November 1985

The Loop Skywire

Looking for an all-band HF antenna that is easy to construct, costs nearly nothing and works great DX? Try this one!

There is one wire antenna that performs exceptionally well on the HF bands, but relatively few amateurs know about it or use it. The purpose of this article is to present what one user has described as the "best kept secret in the amateur circle."

The Loop Skywire antenna is simple and easy to construct, costs nearly nothing, and eliminates the need for multiple antennas to cover the HF bands. It is made of only wire and coaxial cable, and often needs no Transmatch. An efficient antenna, it is effectively omnidirectional over most real earth, and exhibits a good signal-to-noise ratio. The antenna operates on all bands (harmonics) above the design or fundamental frequency and fits on almost every amateur's lot. It also works DX better than any other antenna I have had in the past.

You're suspicious? No antenna does all that? Since 1957, I have used this antenna in many locations with great success every time. There is, of course, no such thing as a "best" antenna. One operator's dream can become another's nightmare. Antennas are very sensitive to their environment. Yet, out of the numerous amateurs I have known who put up this Skywire, not one took it down because of poor performance. Invariably, other antennas, including beams, were

dismantled when the Skywire became their main antenna.

It is curious that many references to this antenna are brief pronouncements that it operates best as a high-angle radiator and is good for only short-distance contacts. Such statements, in effect, dismiss this antenna as useless for most amateur work. This is not the case!

The Antenna

It is quite possible that the Loop Skywire has not been fully studied, analyzed and researched. Those who are able and curious should investigate the polarization of this one. This article does not offer a technical explanation of its performance or operation. Rather, it is a description of

the antenna accompanied by construction hints and actual user comments. Take some time to erect the Skywire and decide for yourself whether it works.

Novices and Extras take note: Here is a simple, single antenna that really works all bands without the need for special stubs or tuning and pruning procedures. A Transmatch in the shack is helpful, but is often unnecessary, especially with tube-final rigs.

The Loop Skywire is a "magnetic" version of the old super SKYBUSTER—the open-wire, center-fed "electric" Zepp that has performed extraordinarily well for many decades. Yet, this one is less difficult to match and use. It can quickly displace that myriad of wires that many have erected in an attempt to work all HF bands. Besides the improve-

> *The Loop Skywire antenna is simple and easy to construct, costs nearly nothing, and eliminates the need for multiple antennas to cover the HF bands.*

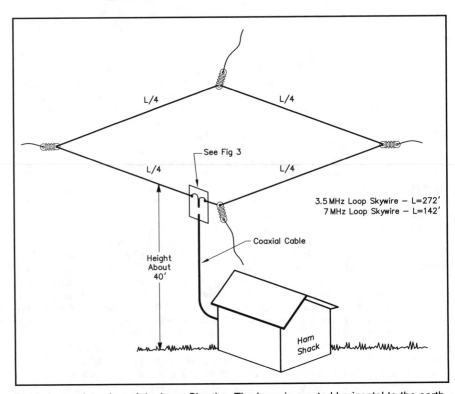

3.5 MHz Loop Skywire – L=272'
7 MHz Loop Skywire – L=142'

Fig 1–A complete view of the Loop Skywire. The Loop is erected horizontal to the earth.

ment in appearance, mutual coupling is greatly reduced. Antennas really do not like neighbors: The more antennas erected, the poorer they all generally work.

The Loop Skywire is shown in Fig. 1. It is simply a loop antenna erected horizontal to the earth. The horizontal position is its secret. Maximum enclosed area within the wire loop is the fundamental rule. The antenna has 1 wavelength of wire in its perimeter at the design or fundamental frequency. If you choose to calculate L_{total} in feet, the following equation should be used:

$$L_{total} = \frac{1005}{f} \qquad \text{(Eq. 1)}$$

where f is frequency in MHz

Given any length of wire, the maximum possible area the antenna can enclose will be with the wire in the shape of a circle. Since it takes an infinite number of skyhooks to hang a circular loop, the square loop (four skyhooks) is the most practical. Reducing the area enclosed by the wire loop further brings the antenna closer to the properties of the folded dipole and both harmonic impedance and feed-line voltage problems can result. Dipole (electric) antennas are only reasonably resonant at their odd harmonics. A little known fact in the amateur community is that loops are reasonably resonant at all harmonics of the design frequency. Loop geometries other than a square are thus possible, but remember the two fundamental requirements for the Loop Skywire: its horizontal position and maximum enclosed area.

Construction

Antenna construction is simple. Generally, a minimum of four skyhooks are required. Fig. 2 shows the placement of the insulators at the loop corners. There are two methods used to attach the insulators: Lock or tie the insulator in place with the loop wire tie shown, or leave the insulator "free" to float or slide along the wire. Most loop users float at least two insulators. This

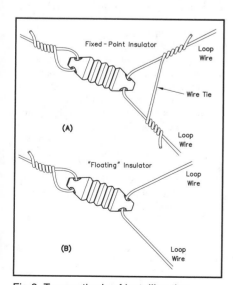

Fig 2—Two methods of installing the insulators at the loop corners.

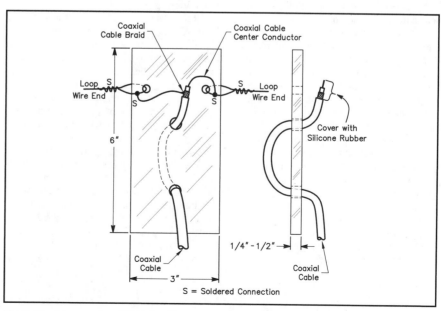

Fig 3—Most users cornerfeed the Skywire. A high-impedance, weather-resistant insulant should be used for the feed-point insulator. Cover the end of the coaxial cable with silicone rubber for protection from the weather and added electrical insulation. Dimensions shown are approximate.

allows pulling the slack out of the loop once it is in the air and eliminates the need to have all the skyhooks exactly placed for proper tension in each leg. I recommend floating two opposite corners. The feed point can be positioned anywhere along the loop that you wish. However, most users corner feed the skywire. Fig. 3 depicts a method of doing this. It is advantageous to keep the feedpoint mechanicals away from the corner support. I usually feed a foot or so off one corner, allowing the feed line to exit more freely. This method keeps the feed line free from the loop support.

If the skyhooks (e.g., trees) move, then at least two of the ropes or guys used to support the insulators should be counter-weighted and allowed to move freely. The feed-line corner is almost always tied down, however. Very little tension is needed to support the loop (far less than that for a di-pole). Thus, counterweights are light weights. Several loops have been con-structed with the use of bungle cords tied to three insulators and the attached ropes tied fast. This eliminates the need for counterweighting.

There is another great advantage to this antenna system. It can be operated as a ver-tical antenna with top-hat loading on all bands as well. This is accomplished by sim-ply keeping the feed-line run from the an-tenna to the shack as vertical as possible and clear of objects. Both feedline conduc-tors are then tied together (via a shorted SO-239 jack, for example), and the antenna is fed against good ground. This method al-lows excellent performance of the 40-meter Loop Skywire on 80 meters, and the 80-meter Loop Skywire on 160 meters. When constructing the loop, connect (solder) the coaxial feed-line ends directly to the loop wire ends. Don't do anything else. Baluns or choke coils at the feed point are not to be used. They are unnecessary. The feed arrangement for operating the loop as a ver-tical antenna is shown in Fig. 4.

Some skeptics have commented that the Loop Skywire is actually a vertical antenna in disguise. Yet when the loops have been used in on-the-air tests with both local and DX stations, on those bands where loop op-eration is possible, the loop operating as a loop consistently "out-signals" the loop operating as a vertical.

Although the loop can be constructed for any band or frequency of operation, the following two Loop Skywires are the star performers. The 30-meter band can also be operated on both.

80-meter Loop Skywire (80-10 meter
loop + 160-meter vertical)
Total Loop perimeter: 272 feet
Square Sidelength: 68 feet

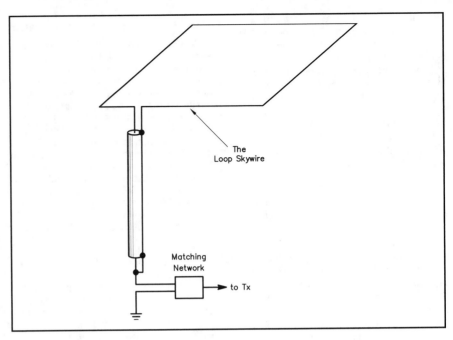

Fig 4—The feed arrangement for operating the loop as a vertical antenna.

40-meter Loop Skywire (40-10 meter
loop + 80-meter vertical)
Total Loop perimeter: 142 feet
Square Sidelength: 35.5 feet

Actual total length can vary from these dimensions by a few feet. Do not worry about tuning and pruning the loop to reso-nance physically. No signal difference was detected on the other end when that method was used. Let the Transmatch do the nec-essary mop up.

Copper wire is usually used in the loop. Lamp or "zip" cord and Copperweld can also be used. Several loops have even been constructed successfully with steel wire, but soldering is difficult.

Recommended height for the antenna is 40 feet or more. The higher the better, especially if you wish to use the loop in the vertical mode. Successful local and DX operation has been reported, however, in several cases with the antenna at 20 feet.

If you are preoccupied with SWR, the reading will depend on your operating fre-quency and the type of feed line used. Co-axial cable is sufficient; open wire does not appear to make the loop perform any better or matching to it any easier. Most users feed with RG-58, RG-59 or RG-62. RG-8 and RG-11 are generally too cumbersome to use. With full power and coaxial cable feeding these loops, feedline problems have not been re-ported to me. The SWR from either of these loops with approximately 100 feet of feed line is rarely over 3:1. For those who understand

SWR, the use of a Transmatch will eliminate all concern for power transfer and maximum signal strength. The SWR in my shack is al-ways 1:1. The highest line SWR usually oc-curs at the second harmonic of the design fre-quency and all other frequencies below that. The Loop Skywire is somewhat more broadbanded than corresponding dipoles, but the loop is efficient: The SWR curves are not "dummy load" flat!

Since the loop is high in the air and has considerable electrical exposure to the elements, proper methods should be employed to eliminate the chance of induced or direct lightning hazard to the shack and operator. Some users simply employ a three connector (PL-259/PL-258/PL-259) weather-protected junction in the feed line outside the shack and completely disconnect the antenna from the rig and shack during periods of possible lightning activity.

The antennas just described are in daily use, and I estimate at least several hundred are working well throughout the world at the present time. Comments offered by some us-ers appear in the accompanying sidebar. They are representative of comments received from the numerous amateurs polled by the author who use the Loop Skywire. There were no dismal or negative results reported with this antenna, which the textbook-touters dismiss as a useless antenna for the HF ham bands. The Loop Skywire is truly a real Loop Skywalker! Try it; you'll like it! I welcome your comments on its use.

CHAPTER FOUR
COLLINEAR ANTENNAS

A 135-Ft Multiband Center-fed Dipole

An 80-m dipole fed with ladder line is a versatile antenna. If you add a wide-range matching network, you have a low-cost antenna system that works well across the entire HF spectrum. Countless hams have used one of these in single-antenna stations and for Field Day operations.

For best results place the antenna as high as you can, and keep the antenna and ladder line clear of metal and other conductive objects. Despite significant SWR on some bands, system losses are low. You can make the dipole horizontal, or you can install it as an inverted V. ARRL staff analyzed a 135-ft dipole at 50 ft above typical

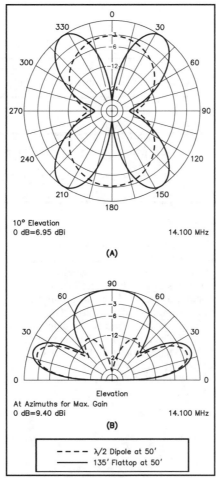

Fig 2—Patterns on 20 m comparing a standard ¹/₂-λ dipole and a multiband 135-ft dipole. Both are mounted horizontally at 50 ft. The azimuth pattern is shown at A, where conductors lie in the 90° to 270° plane. The elevation pattern is shown at B. The longer antenna has four azimuthal lobes, centered at 35°,145°, 215°, and 325°. Each is about 2 dB stronger than the main lobes of the ¹/₂-λ dipole. The elevation pattern of the 135-ft dipole is for one of the four maximum-gain azimuth lobes, while the elevation pattern for the ¹/₂-λ dipole is for the 0° azimuthal point.

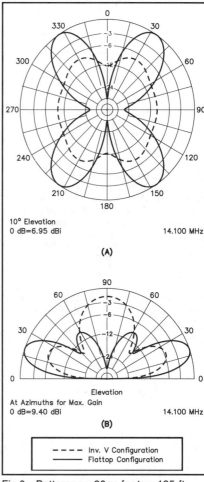

Fig 3—Patterns on 20 m for two 135-ft dipoles. One is mounted horizontally as a flat-top and the other as an inverted V with 120° included angle between the two legs. The azimuth pattern is shown at A, and the elevation pattern is shown at B. The inverted V has about 6 dB less gain at the peak azimuths, but has a more uniform, almost omnidirectional azimuthal pattern. In the elevation plane, the inverted V has a fat lobe overhead, making it a somewhat better antenna for local communication, but not quite so good for DX contacts at low elevation angles.

Fig 1—Patterns on 80 m for 135-ft center-fed dipole erected as a horizontal dipole at 50 ft, and as an inverted V with the center at 50 ft and the ends at 10 ft. The azimuth pattern is shown at A, where conductor lies in the 90° to 270° plane. The elevation pattern is shown at B, where conductor comes out of paper at right angle. At the fundamental frequency the patterns are not markedly different.

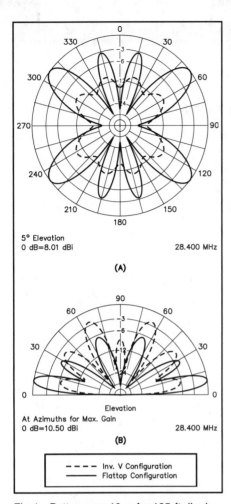

5° Elevation
0 dB=8.01 dBi   28.400 MHz

(A)

Elevation
At Azimuths for Max. Gain
0 dB=10.50 dBi   28.400 MHz

(B)

- - - - Inv. V Configuration
———— Flattop Configuration

Fig 4—Patterns on 10 m for 135-ft dipole mounted horizontally and as an inverted V, as in Fig 20.20. The azimuth pattern is shown at A, and the elevation pattern is shown at B. Once again, the inverted-V configuration yields a more omni-directional pattern, but at the expense of almost 8 dB less gain than the flat-top configuration at its strongest lobes.

Center-Fed Antennas

The simplest and most flexible (and also least expensive) all-band antennas are those using open-wire parallel-conductor feeders to the center of the antenna, as shown in Fig A. Because each half of the flat top is the same length, the feeder currents will be balanced at all frequencies unless, of course, unbalance is introduced by one half of the antenna being closer to ground (or a grounded object) than the other. For best results and to maintain feed-current balance, the feeder should run away at right angles to the antenna, preferably for at least $1/4$ λ.

Center feed is not only more desirable than end feed because of inherently better balance, but generally also results in a lower standing wave ratio on the transmission line, provided a parallel-conductor line having a characteristic impedance of 450 to 600 ohms is used. TV-type open-wire line is satisfactory for all but possibly high power installations (over 500 watts), where heavier wire and wider spacing is desirable to handle the larger currents and voltages.

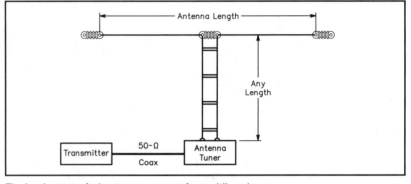

Fig A—A center-fed antenna system for multiband use.

A low-cost antenna system that works well across the entire HF spectrum.

ground and compared that to an inverted V with the center at 50 ft, and the ends at 10 ft. The results show that on the 80-m band, it won't make much difference which configuration you choose. (See Fig 1.) The inverted V exhibits additional losses because of its proximity to ground.

Fig 2 shows a comparison between a 20-m flat-top dipole and the 135-ft flat-top dipole when both are placed at 50 ft above ground. At a 10° elevation angle the 135-ft dipole has a gain advantage. This advantage comes at the cost of two deep, but narrow, nulls that are broadside to the wire.

Fig 3 compares the 135-ft dipole to the inverted-V configuration of the same antenna on 14.1 MHz. Notice that the inverted-V pattern is essentially omnidirectional. That comes at the cost of gain, which is less than that for a horizontal flat-top dipole.

As expected, patterns become more complicated at 28.4 MHz. As you can see in Fig 4, the inverted V has the advantage of a pattern with slight nulls, but with reduced gain compared to the flat-top configuration.

Installed horizontally, or as an inverted V, the 135-ft center-fed dipole is a simple antenna that works well from 3.5 to 30 MHz. Bandswitching is handled by a Transmatch that is located near your operating position.

By John J. Reh, K7KGP   From *QST*, December 1987

An Extended Double Zepp Antenna for 12 Meters

Got a little over 50 feet of horizontal space to spare for a 24-MHz skywire? This simple antenna will beat your half-wave dipole by about 3 dB—and you can phase two of them for even more gain and directivity

According to *The ARRL Antenna Book*, Zepp—short for Zeppelin—is a term long applied to just about any resonant antenna end-fed by a two-wire transmission line.[1] A bit further on in the *Antenna Book*,

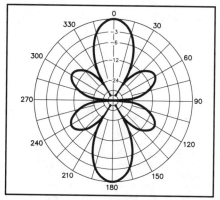

Fig 1–The extended double Zepp antenna consists of two 0.64-λ elements fed in phase.

Fig 2—Horizontal directivity pattern for an extended double Zepp antenna in free space. Relative to a half-wave dipole, it exhibits a gain of approximately 3 dB. The antenna elements lie along the 90°-270° line.

there's a discussion of the extended double Zepp (EDZ) antenna.[2] This interested me because I have always been intrigued by "old-fashioned" wire antennas—and because the old-fashioned extended double Zepp's 3-dB gain over a half-wave dipole would provide performance quite suitable for modern times! The EDZ antenna consists of two collinear 0.64-λ elements fed in phase. Fig 1 shows current distribution in an EDZ, and Fig 2 shows the EDZ's hori-

zontal directivity pattern in free space.

The extended double Zepp's theoretical performance looked good to me, so I designed and built an EDZ antenna for the 12-meter band. Fig 3 shows its configuration. I decided to cut mine for 24.950 MHz. Each EDZ element is 25 feet, 3 inches long, and consists of no. 14 stranded copper wire. The antenna elements are center-fed by a short matching section made of a 5-foot, 5-inch length of 450-Ω open-wire line. Connec-

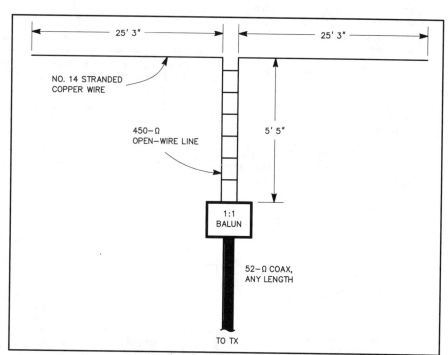

Fig 3–The extended double Zepp at K7KGP, cut for 24.950 MHz. The 450-Ω matching section transforms the EDZ's calculated input impedance (142–*j*555 Ω) to 55 Ω (measured) for connection to 52-Ω coaxial cable by means of a 1:1 balun. The electrical length of the matching section is 52°; the linear dimension shown in the drawing assumes 450-Ω line with a velocity factor of 0.95.

tion to 52-Ω coaxial feed line is made by means of a 1:1 balun transformer. My EDZ is strung between two trees, 35 feet above ground.

Matching Section

Perhaps I am "reinventing the wheel," but I have not seen this matching method elsewhere.[3] The open-wire-line matching section is 52 electrical degrees long (0.145 λ). The matching section transforms the EDZ's input impedance to about 55 ohms, as measured with a noise bridge. The matching-section dimension given in Fig 3 assumes a velocity factor of 0.95 for the 450-Ω line.

Trimming the matching section to size is the only adjustment necessary with the EDZ. Make the transformer a little long to begin with, and shorten it an inch or two at a time to bring the system into resonance. (You can check resonance with a noise bridge or by monitoring the SWR.) Do not change the length of the elements—the EDZ's gain and directivity depend on its elements being 0.64 λ long.

Phasing Two EDZs for More Gain and Directivity

Properly phased, two extended double Zepp antennas can give improved gain and directivity over a single EDZ. Fig 4 compares the calculated horizontal directivity patterns of a single EDZ and an array consisting of two EDZs spaced at 1/8 λ and fed 180° out of phase. Fig 5 compares the vertical radiation patterns of the single and phased EDZs.

Fig 6 shows the dimensions of a practical two-EDZ configuration. With proper ad-

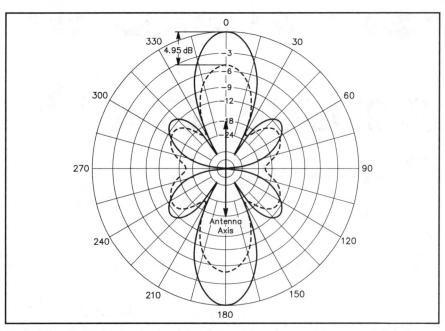

Fig 4—Comparison of calculated horizontal directivity patterns of one extended double Zepp (dotted line) and two EDZs spaced at 1/8 λ and fed 180° out of phase (solid line). The antenna axes lie along the 0°-180° line, and the antennas are mounted 35 feet above average earth. The phased EDZs exhibit nearly 5 dB gain over a single EDZ. This is 7 to 8 dB gain over a half-wave dipole. Beamwidth of the two-EDZ array is 30°. The antenna axis is the same for the single EDZ and both EDZs in the phased array. The two-EDZ configuration characterized here is an end-fire array because maximum radiation occurs along its axis.

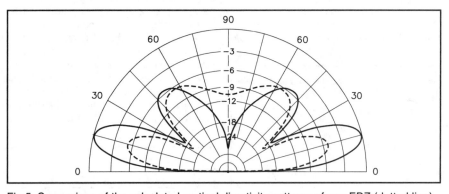

Fig 5—Comparison of the calculated vertical directivity patterns of one EDZ (dotted line), and two EDZs spaced at 1/8 λ and fed 180° out of phase (solid line). The antenna axis lies along the 0° line.

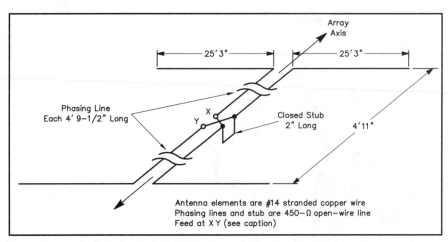

Fig 6—One method of phasing two EDZs for greater gain and directivity. The array is bidirectional, with maximum radiation occurring along the array axis. The impedance across points X and Y is 50 Ω, balanced; with a 1:1 balun at XY, the array can be fed by means of 52-Ω coaxial cable. The stub, 1.5° long, cancels a capacitive reactance of approximately 13.5 Ω at the feed point. This array works well, but its matching system is clumsy because the combined length of the phasing lines is greater than the spacing of the two EDZs. Fig 7 shows a proposed feed method that takes up less space.

Properly phased, two extended double Zepp antennas can give improved gain and directivity over a single EDZ.

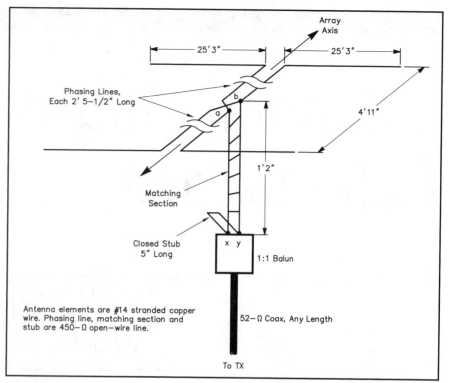

Fig 7–Proposed alternative method of phasing two EDZs. In this arrangement, the length of each phasing line is half the EDZ spacing. Calculated impedance across points a and b is 15–j112 Ω. The matching section–11° in length–transforms this to a calculated impedance of approximately 55–j32 Ω (balanced) across points x and y. The stub, 4° long, cancels the capacitive reactance (32 Ω). A 1:1 balun transformer allows the array to be fed by means of a 52-Ω coaxial cable. See text.

justment, it exhibits an SWR of 1.3:1 across the 24-MHz band. In the array I built, lightweight broom handles serve as spreaders between the element ends; the center spreader is a wooden slat. I used nylon rope to haul the array up between two trees. This antenna system works well, but poor propagation has precluded a thorough tryout so far. The contacts I have had with it have been entirely satisfactory.

The matching method shown in Fig 6 is somewhat clumsy because the combined length of the phasing lines is greater than the spacing between the EDZs. The feed method shown in Fig 7 should be easier to build because the combined length of the phasing lines equals the spacing between the EDZs. I have not tried this matching method, but I'm confident that my calculated dimensions are close to what will actually be encountered.[4]

Conclusion

If the extended double Zepp has caught your attention, but 12 meters hasn't, you can scale the linear dimensions given here for other bands of interest. Once your EDZ is up and working, I think you'll agree that the performance of the "old-fashioned" extended double Zepp isn't old-fashioned at all!

Notes

[1] *The ARRL Antenna Book*, J. Hall, ed. (Newington: ARRL, 1984), p 5-4.
[2] *The ARRL Antenna Book*, p 6-8.
[3] K7KGP's matching technique is a "reinvention

Trimming the matching section to size is the only adjustment necessary with the EDZ.

of the wheel" of which he can be proud. Termed the series section transformer, it appears in *The ARRL Antenna Book* and *The ARRL Handbook*. The series section material in these books is based on Frank A. Regier, "Series-Section Transmission-Line Impedance Matching," *QST*, Jul 1978, pp 14-16.—*Ed*.

[4] K7KGP's calculations were confirmed by Rus Healy, NJ2L, of the ARRL HQ Technical staff using the Smith Chart and the Mini-Numerical Electronics Code (*MININEC*) on an IBM personal computer. Data for the plots in Figs 4 and 5 were also generated by means of MININEC.—*Ed*.

Extended Double-zepp Calculations

I found the 3-dB gain figure attractive. But I wanted to build the antenna for 20 meters instead of 12 meters.

John Reh's article, "An Extended Double Zepp Antenna for 12 Meters," is interesting from both a technical and constructional point of view; I found the 3-dB gain figure attractive. But I wanted to build the antenna for 20 meters instead of 12 meters. Using John's article and performing some research in the *ARRL Handbook*, I reworked the calculations and came up with the following formulas. I thought other *QST* readers who wanted to build EDZs for other frequencies might find this information useful.

- A constant (984) is used to determine the electrical length of a wire in feet: W(ft) = 984/f, where f is the desired operating frequency in MHz.
- The overall length of each leg is calculated by

$$L = W \times 0.64 \qquad \text{(Eq 1)}$$

- The 450-ohm openwire matching line for a single EDZ is calculated by

$$M(sgl) = 52/360 \times 0.95 \times W \qquad \text{(Eq 2)}$$

The 450-ohm line is made of no. 18 wire spaced 1-inch center to center. The line has a velocity factor of 0.95. For phased EDZs, the following calculations apply:

- The length of the 11-degree matching line is calculated by

$$M(pha) = 11/360 \times W \times 0.95 \qquad \text{(Eq 3)}$$

- The length of the 4-degree matching stub is calculated by

$$S(pha) = 4/360 \times W \times 0.95 \qquad \text{(Eq 4)}$$

- Spacing for phased EDZs is W/8.

The input impedance of the dipole (142–*j*555) was calculated for no. 14 wire. The *ARRL Handbook* defines the characteristic impedance of no. 12-14 wire as 500-600 ohms at a height of 10 to 30 feet. If you use other wire sizes for the dipole itself, the matching section length may require changing, so start with a longer matching section and trim it as required.

For a 20-meter EDZ, the numbers worked out this way: The antenna has an overall length of 88 feet and a 9.53-ft matching section. A 10-m phased EDZ looks quite easy to build, and should exhibit a gain of 8 dB over a dipole. I'm not sure if this design approach will work for 2-meter band antennas, but the dimensions look quite manageable.—*Bob Mandeville, N1EDM*

By Walter E. Salmon, VK2SA

From *QST*, October 1955

The "Extended Lazy H" Antenna

Rotary beams were unknown in the early days of Amateur Radio, and most hams contented themselves with horizontal or vertical wires from which, after much patient work, they obtained varying degrees of effectiveness. With the development of the Yagi antenna the 2, 3 and 4-element rotary beam became commonplace, and it would appear that the trend in this direction is increasing, particularly with amateurs residing in thickly populated areas where land space is limited. No comment will be included about V beams and rhombics, since this article is written for the amateur who, although he is interested in operating on several bands, is not prepared to erect a costly mast structure to support several beams and also does not have the relatively unlimited space necessary for the usual "dream" antenna farm.

The antenna to be described is completely original and to the writer's knowledge has not been described in any local or overseas journal. We have "ZL Speeials" and "G8PO antennas" and, for want of a name, this antenna might be called the "extended lazy H." Several years ago a conventional lazy H antenna was cut for 14 MHz and installed at VK2SA. This aerial consisted of two horizontal collinear elements stacked and separated a half wavelength. The top of the array was supported by two 41-foot masts, thus leaving the bottom section only 9 feet above the ground. The effective height of this type of antenna is measured from the halfway point between top and bottom elements and thus, in this case, the effective height was about 25 feet. The observed effectiveness was only about equal to a full-wave Zepp 41 feet high.

Attention was then directed to the possibilities of the "extended double Zepp" described in *QST* for June, 1938. The height of one mast was increased to 45 feet, to compensate for ground slope, and the antenna was cut for 14 MHz and erected for NE-SW directivity. Improved effectiveness by comparison with the full-wave antenna was apparent on 14-MHz W contacts. In addition, some excellent phone contacts were made on 7 MHz with W stations. Results on 21 MHz indicated a number of major lobes that gave good DX coutacts.

From the results it would appear that this type of antenna possesses the desirable feature of good effectiveness on several amateur bands. The gain of the extended double Zepp is given in most textbooks as 3 dB.

The theoretical gain of the conventional lazy H antenna is given as close to 6 dB, but it was considered attainable only if it could be supported about 70 feet in the air, so that the bottom elements were at least a half wavelength above ground. This was impossible with the existing masts. Consideration was then given to the possibility of adding two additional extended halfwave lower elements to the extended double Zepp. The additional elements were connected 21 feet down on the feed line,[1] as shown in Fig. 1 and the feed line was transposed to give the proper phasing.

Results with the modified antenna were very gratifying, as was the ability to

[1]The point 21 feet down the feed line is a voltage loop, and one would normally connect half-wavelength elements at this point for in-phase drive of all elements. The modification by VK2SA is not the simplest array to analyze, but in view of his excellent results it is thought to be of considerable interest.—*Ed.*

operate readily on three bands with the one antenna system. Although the directional characteristics on 21 MHz are not yet known completely, the signal reports indicate the presence of major lobes giving good general coverage. On 7, 14 and 21 MHz an antenna tuner is used, and an open-wire line with 4-inch spacing is used between tuner and antenna.

On 14 MHz the antenna has outperformed all previous wire antennas tried out for W contacts on both long and short paths. The lower two elements were added to the extended double Zepp on December 19, 1954, and numerous W phone contacts have been made since that date. The majority of the signal reports are S8 and S9, and nothing below S6 from East Africa. The power input to the transmitter is 75 watts.

An analysis of all signal reports indicates equal if not better performance compared with rotary beams, and it would appear that the gain exceeds 6 dB. Comparison reports have also been made by the simple expedient of removing the two lower elements—the antenna then becomes an extended double Zepp—and the signal was reported to drop 2 and sometimes 3 S points.

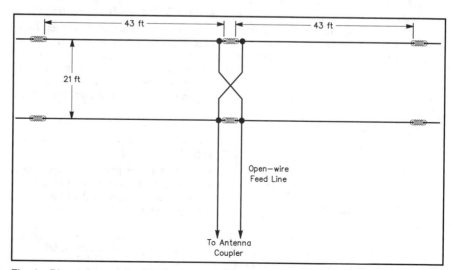

Fig. 1—Dimensions of the "VK Special" 7, 14 and 21-MHz beam antenna of VK2SA. Whether the antenna coupler will be series- or parallel-tuned will depend upon the length of the feed line and the band in use. At VK2SA the upper wire is 40 feet above the ground.

By Douglas J. Fouts, KI6QR

From *QST*, March 1989

Collinear Phased Antennas for the HF Bands

Need a good-performing wire antenna? A collinear phased array could be just what you're looking for!

The thrill of chasing DX is an integral part of Amateur Radio. With the increase in sunspot activity, there's more DX fun every day! Unfortunately, newcomers (and some old-timers) sometimes think that the only way to work DX is with the aid of costly towers, rotators and beams. This article shows you how to design and build an effective DX antenna called a collinear phased array. Collinear antennas are easy and inexpensive to construct, and if you already have a dipole or an inverted V, you already have many of the materials you'll need.

Collinear-Array Fundamentals

Basically, a collinear array is nothing more than two or more radiating elements that are strung together, end to end.[1] The radiating elements are usually a half wavelength long, but other lengths can be used. Fig 1 shows a collinear array that is composed of three half-wave, center-fed dipoles. If the three transmission lines that lead from the common feed point in Fig 1 to the centers of each dipole are the same length, the currents in all three dipoles will always be traveling in the same direction

at a given time (the three dipoles are in phase). The term collinear is used because the radiating elements are arranged geometrically in a straight line.

If the three dipoles shown in Fig 1 are fed in phase, the fields created by the dipoles will also be in phase. At a distant receiving antenna in a direction perpendicular to the array elements, the energy radiated by the three dipoles adds, creating a received signal that is stronger than if only one dipole was used. This principle also applies to the reception of incoming waves: A perpendicular incoming wave strikes all three dipoles at the same time. The signals induced in the dipoles add in phase, making the received signal stronger than if only one dipole was used. The theoretical gain of a collinear phased array over a half-wave dipole is 1.9 dB for an array with two elements, 3.2 dB for a three-element antenna and 4.3 dB for an array with four elements.

Feeding Collinear-Phased Arrays

The use of multiple feed lines as shown in Fig 1 has the advantage of ease of construction and tuning. The only things you need to tune this antenna system are a tape measure, a transmitter and an SWR bridge. The use of multiple feed lines has several

disadvantages, however. Feed lines should run perpendicular to the antenna for at least a half wavelength before turning any corners. This can be an unsightly mess, difficult to implement (depending on the installation) and, most of all, costly. Table 1 shows the minimum amount of parallel conductor transmission line that is necessary to connect multiple elements to a common feed point.

The preferred methods of feeding collinear phased arrays are shown in Figs 2 and 3. Fig 2 shows the feed system for an antenna with an odd number of elements, and Fig 3 is that for an antenna with an even number of elements. The advantage of these feed methods is that only one feed line is required, and its length isn't critical. In general, the feed methods shown in Figs 2 and 3 are less costly, easier to construct, and visually unobtrusive. The disadvantage is that some means (such as an RF noise bridge) is required to adjust the quarter-wave phasing lines.

The short arrows next to the antenna elements in Figs 2 and 3 indicate the direction of current flow during half of a cycle of the applied signal. During the following half cycle, all currents flow in the reverse direction. In a long, continuous wire, the direction of current flow reverses every half

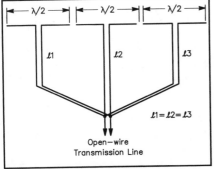

Fig 1–A three-element, collinear phased array fed with equal-length sections of open-wire line. (This is the simplest way to feed a collinear array.) See text.

Table 1
Number of wavelengths of open-wire transmission line required to connect two to five elements to a common feed point

Number of elements	Wavelengths of transmission line
2	1.5
3	3.0
4	5.0
5	7.5

Newcomers (and some old-timers) sometimes think that the only way to work DX is with the aid of costly towers, rotators and beams.

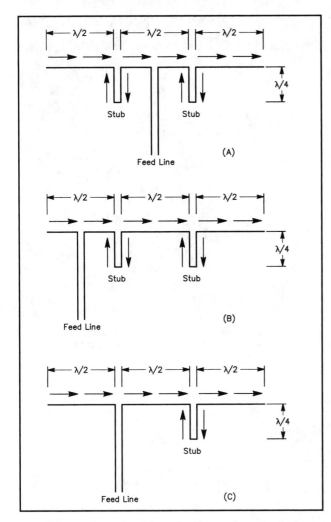

Fig 2–Feeding an odd number of collinear elements. At A, balanced current feed. At B, unbalanced voltage feed. At C, current feed.

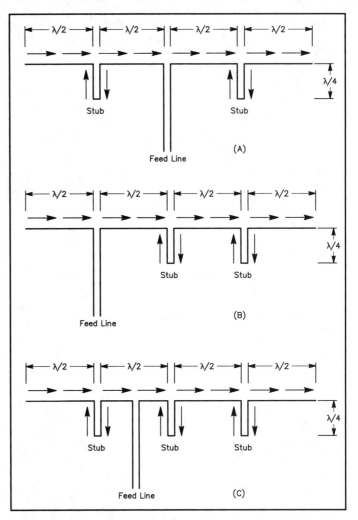

Fig 3–Feeding an even number of collinear elements. At A, balanced voltage feed. At B, unbalanced voltage feed. At C, current feed.

wavelength. Figs 2 and 3 show that all currents in the radiating sections are in phase.

The short vertical sections shown in Figs 2 and 3 are quarter-wave phasing stubs. The current flowing on one side of a stub is equal in amplitude to the current flowing on the other side at every point along the length of the stub. Because the currents flow in opposite directions, their magnetic fields cancel, and the stub does not radiate. The current amplitude is not constant along the length of the stub: The current is at a minimum at the top of the stub, and reaches a maximum at the bottom of the stub.

Basically, a collinear array is nothing more than two or more radiating elements that are strung together, end to end.

Both voltage- and current-feed methods are shown in Figs 2 and 3. Voltage feed is so named because the antenna is fed at a point of maximum voltage (minimum current). Current feed is feed at a point of maximum current (minimum voltage).

Collinear phased arrays work best if they

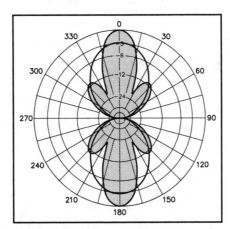

Fig 4–Radiation patterns of center-fed collinear phased arrays. The unshaded pattern is for an antenna with two elements; the shaded pattern is for a four-element antenna.

are balanced (equal signal amplitudes flow on both sides of the antenna at a given time). Therefore, current feed is preferred for an antenna with an odd number of elements (Fig 2A). The feedpoint impedance of a current-fed antenna is slightly over 300 Ω and provides a reasonable match to 300- or 450-Ω openwire transmission line. Voltage feed is preferred for an antenna with an even number of elements (Fig 3A). The feedpoint impedance of a voltage-fed antenna is over 1000 Ω. If a particular installation requires an off-center feed, current feeding is the preferred method because of its more favorable feed-point impedance.

Radiation Patterns

Collinear phased arrays are bidirectional. In addition to some minor lobes, they produce two major lobes perpendicular to the antenna wire. The more elements the antenna has, the more narrow the major lobes (and the more numerous the minor lobes). Fig 4 shows the radiation pattern for collinear phased arrays with two and four elements. In addition to the coverage provided by the major lobes, the minor lobes usually provide adequate coverage for local communication.

Construction

The first step in constructing a collinear phased array is to decide the directions of the major lobes and how many elements the antenna should have. This depends on the location of the desired DX, whether or not these locations all lie in (or close to) a straight line and how much room there is at the installation site. If the desired DX locations all lie in (or close to) the same direction (or 180° from each other), the antenna can be built with many elements, assuming that the necessary space is available. On the other hand, if the desired DX is fairly spread out, the wider major lobes of an antenna with fewer elements may be more desirable.

If the gain and directivity of a three-element array is desirable but there isn't enough space for it, and if there is more than the required amount of space for a two-element array, an extended double Zepp (EDZ) should be considered. The EDZ—a special case of the two-element collinear phased array—has a gain of approximately 3 dB over a half-wave dipole, and is only 1.28 λ long. For information on this antenna, see *The ARRL Antenna Book* and *QST*.[2]

Collinear phased arrays work best if they are erected parallel to the ground. This means that a support is needed at each end—the higher the support, the better. If minimizing cost is important, or if less directivity is desired, the antenna can be erected as an inverted V. If this is done, the apex angle should not be less than 90°. The antenna may have to be raised and lowered several times during construction and tuning, so it is a good idea to use support structures with pulleys, as shown in Fig 5. The antenna can then be raised and lowered easily with halyards.

Materials

The radiating sections of the array should be constructed from no. 12 or 14 copper-clad steel wire, such as Copperweld. The extra strength of steel wire is desirable because the antenna must bear the weight of the phasing stubs and transmission line. If minimizing cost is important, the antenna can be built from no. 12 insulated copper wire (the kind used for house wiring). If copper wire is used, however, the antenna may not stay up during severe weather, and the conductors may stretch excessively if they're not well supported.

The radiating sections of a collinear phased array must be an electrical half wavelength long, which means that they will be slightly shorter than the calculated value given by the equation

$$\text{length (feet)} = 468 \div \text{f (MHz)} \qquad \text{(Eq 1)}$$

There are two methods that can be used to find the correct electrical length for the materials being used; both start with the same step. Calculate the length of a half wavelength of wire using Eq 1. Use a frequency at the center of the desired band for this calculation.

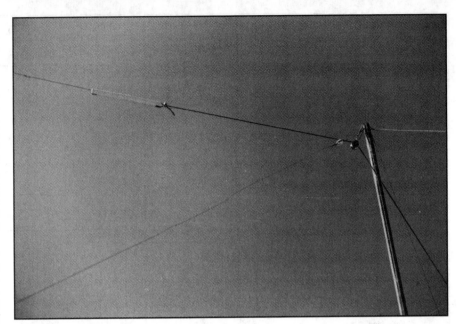

Fig 5—Supporting mast for a collinear phased array. The guy cable that pulls in the same direction as the antenna prevents the mast from bending backward when tension on the antenna is released.

Fig 6—A coiled quarter-wave phasing stub constructed from no. 12 insulated copper wire.

The first step in constructing a collinear phased array is to decide the directions of the major lobes and how many elements the antenna should have.

The next step in using the first method is the construction of a half-wave dipole that uses the same antenna wire and insulators that will be used in the construction of the collinear antenna. Connect a 50- or 75-Ω transmission line to the dipole, and raise the antenna into position. Good-quality transmission line is not necessary for this test; an old piece of RG-58 or RG-59 will work fine. A balun is not required either. Resonate the dipole by pruning equal amounts off each end for best SWR. Once the dipole is resonant, it can be lowered and measured to determine the correct length for the half-wavelength elements.

The second method for determining the proper element length is similar to the first, except that you'll need to build a fullwave dipole and then prune it for resonance at half of the desired center frequency. (You can only use this method if half of the desired center frequency falls in an amateur band.)

The first method is preferred if you are building a current-fed antenna, because the test dipole can be used as an element of the collinear array. The second is the preferred method when constructing a voltage-fed antenna, for the same reason. Once an electrical half wavelength has been determined, the radiating elements for the collinear antenna can be cut and assembled. A word of caution is in order here: The ends of the radiating elements are points of high voltage. If you intend to use high power (1 kW or more), large insulators, such as those shown in Fig 5, should be used. If large insulators aren't available, or if cost must be minimized, two smaller insulators in series can be substituted.

The Phasing Stubs

The quarter-wave phasing stubs can be

constructed from almost any type of wire. If the antenna is to be fed with less than 100 W, 300-Ω TV twin lead can be used. If higher power levels are anticipated, no. 14 conductor, 1-inch spaced, balanced transmission line can be used. Widespaced line such as this can, however, be hard to resonate because of its low Q. It is also more expensive than no. 12 insulated copper wire.

The phasing stub shown in Fig 6 is constructed from no. 12 insulated-copper house wire: Two quarter-wavelength pieces of wire are twisted tightly together. No. 14 zip cord could also be used to make the phasing stubs. Figs 2 and 3 show the phasing stubs hanging straight down, but this is visually obtrusive. (To make the stubs even less noticeable, make them of wire with blue or clear insulation.) Because the stubs don't radiate, they can be coiled, as shown in Fig 6.[3] This lowers their resonant frequency a bit, so coil them before pruning them to resonance.

To resonate the phasing stubs, an RF noise bridge is required. Any of the commonly available commercial units work fine for this.[4] Connect the bridge's unknown input to the end of the phasing stub that will eventually connect to the radiating elements. Leave the other end of the stub open. An open quarter-wave transmission line behaves like a series resonant circuit. Follow the instructions that come with your bridge for finding the resonant frequency of a series resonant circuit. The actual resonant frequency should be lower than the calculated value. Prune each stub until it is resonant at the antenna's center frequency, and then attach it to the radiating elements. Solder together the wires at the other end of the stub.

Collinear phased arrays should be fed with balanced transmission line. At output power levels of 100 W or less, if the line length is not too great, 300-Ω TV twin lead can be used. TV twin lead is too lossy for long runs, though, and it can't handle high power. For runs of more than 1 λ and/or high-power applications, transmission line with no. 14 (or heavier) conductors should be used. If this is unavailable, or if cost is being kept to a minimum, a suitable transmission line can be made from two lengths of no. 12 or 14 copper wire. Keep the wires 1 inch apart with plastic-sheet or weatherproofed-wood spacers attached at 6-inch intervals along the length of the line.

If the antenna is current fed and has been designed and built correctly, the transmis-

sion-line SWR should be less than 3:1. The SWR on a voltage-fed antenna will be higher. With a short, lowloss transmission line, a high SWR can be tolerated. If, however, the line length is more than 1 λ and low-loss line isn't used, a matching stub may be required. Information on stub matching with transmission lines can be found in *The ARRL Antenna Book*.[5]

Because of the SWR-protected final amplifiers in today's solid-state, broadband rigs, antenna tuners are needed with most antennas—including collinear phased arrays—to present a suitable SWR to such radios. If you use a tuner that doesn't have a balanced-line output, connect the feed line to a 4:1 balun and run a short piece of coax from the balun to the tuner. If a tuner is used, a collinear array can be used on bands other than the one that it is designed for, with some sacrifice in performance.

A Collinear-Phased Array in Action

I constructed my antenna for 21.225 MHz, the center of the 15-meter band. The antenna has four elements, is voltage fed through a home-brew L-network tuner, 50 feet of RG-213 coax, a 4:1 balun and 20 feet of open-wire transmission line. The antenna height is 30 feet, and the wire is oriented in a southwest/northeast direction. The major lobes point toward the middle of South America and Japan.

I tested the antenna during the 1988 CQ WPX SSB Contest, in which I worked many stations in South America and Asia. Signal reports were quite good. After the contest, I had lengthy QSOs with several stations in South America and Japan, during which I received signal reports indicating that the collinear works about as well as a three-element triband Yagi. I also made some contacts with stations in Eastern Europe, Australia and New Zealand. Although signal levels off its ends are lower, the antenna is not totally useless

in these directions.

With the aid of a tuner, my 15-meter collinear is useful on other bands. I used the antenna on 40 and 80 meters during the WPX contest, and I worked stations all over the US with my 100-W signal.

Conclusion

After constructing and using the 15-meter collinear antenna as described in this article, I have come to the conclusion that, for my needs, it is probably the best performing antenna for the money. Like many hams, when I first got started in ham radio I put up a multiband dipole and fed it through a tuner. Because of this exercise, two guyed masts, antenna wire, transmission line, a balun, a tuner and some insulators were already on hand. I borrowed an RF noise bridge, so the only parts I had to buy for this project were the wire for the phasing stubs and a few insulators. In total, I spent less than $10—a lot less than I'd have to spend to put up a small Yagi!

Acknowledgment

I thank Dr Stephen I. Long, AC6T, for loaning me his RF noise bridge.

Notes

[1] For more information on collinear-array theory, see G. L. Hall, ed, *The ARRL Antenna Book*, 15th edition (Newington: ARRL, 1988).

[2] See note 1 and J. Reh, "An Extended Double Zepp Antenna for 12 Meters," *QST*, Dec 1987, pp 25-27. Also see J. Reh, "Extended Double Zepp Calculations," *QST*, Aug 1988, p 51.

[3] At any point along the stub, most of the energy is concentrated close to and between the conductors. Therefore, coiling the stubs does not degrade their performance. See J. D. Kraus and K. R. Carver, *Electromagnetics*, 2nd edition (New York: McGraw-Hill, 1973), pp 520-521.

[4] Palomar Engineers, MFJ and Heathkit all make RF noise bridges.

[5] See note 1, Chapters 26 and 28.

> *If the desired DX locations all lie in (or close to) the same direction (or 180° from each other), the antenna can be built with many elements, assuming that the necessary space is available.*

CHAPTER FIVE

V AND RHOMBIC ANTENNAS

"V" Antennas

Simple and combination forms of long-wire directive systems.

The lobe of maximum radiation from a single long wire makes a more acute angle with the wire, and the power in the lobe becomes greater, as the length of the wire (in wavelengths) is increased. There are several ways in which long harmonic wires ean be combined to add the effects of the single wires and give greater gain.

The Echelon Antenna

If two long wires are made parallel to each other and excited out-of-phase, the radiation will cancel in the direction perpendicular to the plane of the wires. The radiation is then maximum in the plane of the wires and at angles equal to those of the lobes for a single wire of equal length. If the two wires are staggered, two of the four lobes will disappear, providing the stagger and spacing of the wires is correct for their length. This form of antenna is shown in Fig. 1 and is called the "echelon" antenna. It is normally used with the wires in the horizontal plane, but it can be used with the wires in a vertical plane. However, when the wires are used in the vertical plane, they must be slanted to make the lobe come down to a reasonable angle with the horizontal and this necessitates supports that are too high for all practical purposes. The horizontal echelon antenna should be at least a wavelength above ground for best results.

The amount of stagger can be calculated from

$$s\,(\text{feet}) = \frac{492 \sin \alpha}{f\,(\text{MHz}) \sin 2\alpha}$$

and the distance between wires is given by

$$d\,(\text{feet}) = \frac{492 \cos \alpha}{f\,(\text{MHz}) \sin 2\alpha}$$

where α = angle of maximum radiation from a single wire.

As an example, an echelon antenna with 3-wavelength legs would be spaced a half-wavelength and staggered 0.29 wavelength.

The echelon antenna can be made uni-directional by introducing another pair of wires at quarter-wave space and 90° phase relation with the first pair, but this is a refinement used only in commercial installations.

The two wires of the echelon should be fed as shown in Fig. 1 so that radiation from the feeders will be minimized and not tend to cancel the directivity.

The principal objection to the echelon antenna is the fact that the gain only becomes significant when wires of at least three wavelengths are used and when the height can be made at least a wavelength. These figures are not hard to attain at 28 MHz and higher, however.

The "V" Antenna

Two wires combined to form a "V" at such an angle that the main lobes reinforce along the line bisecting the V make a very effective directional antenna. If the two sides of the V are excited 180° out-of-phase, by connecting the two-wire feed line to the apex of the V, the lobes add up along the line of the bisector and tend to cancel in other directions, as shown in Fig. 2. The V antenna is essentially a bi-directional system, and the gain depends upon the length (in wavelengths) of the wires. The V is a simple antenna to build and operate, providing the necessary room is available, and with tuned feeders it can be operated satisfactorily on several bands, although it is of course optimum for only one. Nevertheless, it will show considerable gain on several bands, the gain increasing as the frequency increases. The longer the V the less will be the departure from optimum angle on several bands.

The chart in Fig. 3 gives the dimensions that should be followed for an optimum design to obtain maximum power gain from a V beam. The wave angle referred to is the vertical angle of maximum radiation for a height of $1/2$ wavelength, and this angle becomes less for any given length as the height above ground is increased. Tilting the whole horizontal plane of the V will tend to increase the low-angle radiation off the low end and decrease it off the high end. If the ground slopes, the antenna should be made parallel to the ground and preferably with the open end of the V down the slope.

The gain of the V beam can be increased by stacking two beams one above the other, a halfwavelength apart, and feeding them so that the legs on one side are in phase with each other and out-of-phase with the legs on the other side. This will result in a greatly lowered angle of radiation. The bottom V should be at least a quarter-wavelength above the ground and preferably a half-wavelength.

Two V beams can be broadsided to form

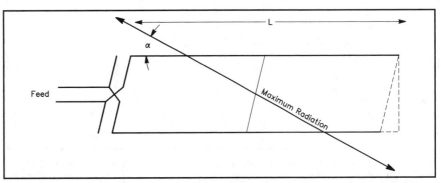

Fig 1–The "echelon" antenna uses two long parallel wires excited out-of-phase. In its simplest form it is a bi-directional affair, although it can be made uni-directional by the addition of two more elements. See text for dimensions "s" and "d." The angle is the angle of the major lobe for the length L.

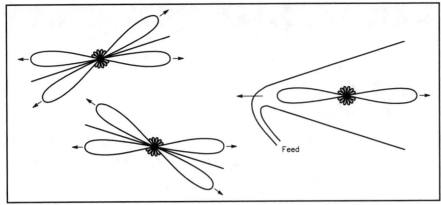

Fig 2—Two long wires and their respective patterns are shown at the left. If these two wires are combined to form a "V" whose angle is twice that of the major lobes of the wires, and the wires are excited out-of-phase, the radiation along the bisector of the V adds and the radiation in the other directions tends to cancel.

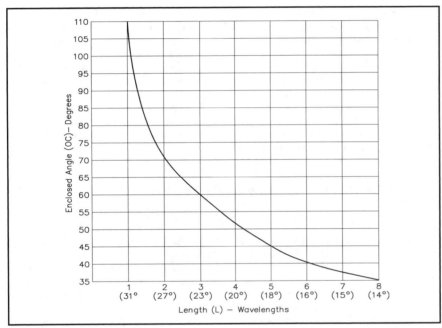

Fig 3—Design chart for horizontal "V" antennas. The enclosed angle between sides is shown plotted against the length of the legs in wavelengths.

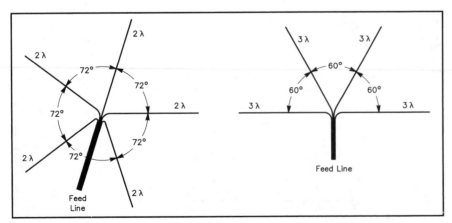

Fig 4—Two suggestions for V-beam combinations. The two-wavelength affair at the left gives good general coverage, because it gives a choice of five directions. The system at the right gives more gain but a choice of only three directions. (Since the systems are bi-directional, "direction" means along a bi-directional line.) Other combinations could be: seven 4-wavelength elements arranged radially at 51½°; five 5-wavelength elements arranged half-radially at 45°; nine 6-wavelength elements arranged radially at 40°.

a "W" and give greater gain. However, two feed lines are required and this fact, plus the five poles required to support the system, renders it normally impractical for the amateur. It is used by many commercial short-wave stations.

The V beam can be made uni-directional by using a reflector at least a quarter-wavelength in back of the antenna. A better plan is to use another V placed an odd quarter-wavelength in back of the first and to excite the two with a phase difference of 90°. The system will be unidirectional in the direction of the antenna with the lagging current for quarter-wave separation. However, the parasitic or driven reflector is not normally employed by amateurs because it restricts the use to one band, although it has proved to be quite effective in commercial work.

The V can be made uni-directional and aperiodic by terminating the open ends of the V to ground through resistors. These resistors must dissipate almost half the power fed to the antenna and the ground connection must be an excellent one. Because of the practical difficulties involved, terminated Vs are not often used, although they present excellent possibilities.

Feeding the V

The V beam is most conveniently fed by tuned feeders, since they permit multi-band operation. If an untuned line is used, the quarter-wave matching section is as convenient as any, since it allows the entire system to be tuned before attaching the feeders, by simply adjusting the shorting bar on the matching section. The length of the wires in a V beam is not at all critical, but it is important that both wires be of the same electrical length. Balanced feeder currents (in a tuned line) give sufficient indication of balanced lengths in the antenna proper.

The terminated V is fed by a 600-ohm line, and a good match will be obtained with almost any combination. The terminating resistors should be adjusted for minimum standing waves on the feeders and the performance should be checked on several frequencies throughout a band.

V-Beam Combinations

The V beam lends itself admirably to a general coverage system by arranging several V beams radially and selecting the one in the desired direction by switching to the proper feeder combination. Antennas of this type have been used at several outstanding DX stations with excellent results. Fig. 4 shows the general principle and gives the length and angle combinations that can be used.

The proper pair of wires can be selected by running all of the feeder wires into the shack (the feeder spacers will be hoops instead of bars) or the antennas can be switched at the pole by means of a remote relay.

In any V-beam combination, it is important that the flat-top lengths be so adjusted that going from one combination to the next will not affect the tuning at the transmitter end. This can be readily done by progressively pruning the antennas at the far ends until they all match up. Care must be taken, of course, that the wire in each feeder is exactly the same length if they are all brought into the station.

Obtuse-Angle Vs

It might be considered that an obtuse-angle V could be used, since if it were fed at one end and properly proportioned the lobes should reinforce along the perpendicular to the bisector. However for the same amount of wire, the obtuse-angle V is definitely inferior to the acute-angle V. Two obtuse-angle λ Vs, side by side to form a diamond are something else again, and require a chapter by themselves.

From *The ARRL Antenna Book*, 18th edition

Lobe Alignment

It is possible to align the lobes from the individual wires with a particular wave angle by the method described in connection with Fig 3. At very low wave angles the required change in the apex angle is extremely small; for example, if the desired wave angle is 5° the apex angles of twice the value given in Fig 1 will not need to be reduced more than a degree or so, even at the longest leg lengths which might be used.

When the legs are long, alignment does not necessarily mean that the greatest signal strength will be obtained at the wave angle for which the apex angle is chosen. Keep in mind that the polarization of the radiated field is the same as that of a plane containing the wire. As illustrated by the diagram of Fig 3, at any wave angle other than zero, the plane containing the wire and passing through the desired wave angle is not horizontal. In the limiting case where the wave angle and the angle of maximum radiation from the wire are the same, the plane is vertical, and the radiation at that wave angle is vertically polarized. At in-between angles the polarization consists of both horizontal and vertical components.

When two wires are combined into a V, the polarization planes have opposite slopes. In the plane bisecting the V, this makes the horizontally polarized components of the two fields add together numerically, but the vertically polarized components are out of phase and cancel completely. As the wave angle is increased, the horizontally polarized components become smaller, so the intensity of horizontally polarized radiation decreases. On the other hand, the vertically polarized components become more intense but always cancel each other. The overall result is that although alignment for a given wave angle will increase the useful radiation at that angle, the wave angle at which maximum radiation occurs (in the direction of the line bisecting the V) is always below the wave angle for which the wires are aligned. As shown by Fig 7, the difference between the apex angles required for optimum alignment of the lobes at wave angles of 0° and 15° is rather small, even when the legs are many wavelengths long.

For long-distance transmission and reception, the lowest possible wave angle usually is the best. Consequently, it is good practice to choose an apex angle between the limits represented by the two curves in Fig 7. The actual wave angle at which the radiation is maximum will depend on the shape of the vertical pattern and the height of the antenna above ground.

When the leg length is small, there is some advantage in reducing the apex angle of the V because this changes the mutual impedance in such a way as to increase the gain of the antenna. For example, the optimum apex angle in the case of 1-λ legs is 90°.

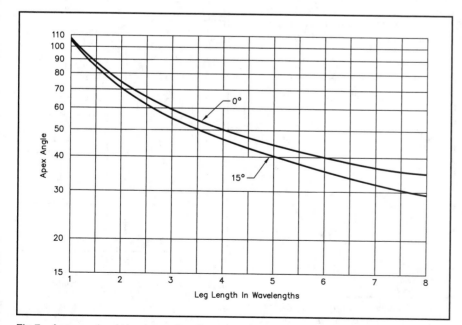

Fig 7—Apex angle of V antenna for alignment of main lobe at different wave angles, as a function of leg length in wavelengths.

Multiple V Beams

High Gain in All Directions with Four Wires

This is a description of the V-beam antenna system used at DL4ZC. Although it is impossible to "design" such a system for more than one band, since the optimum angle varies with frequency and leg length, experience has shown that such a combination of long wires at a compromise angle can be quite effective as a multiband antenna. For those who have the necessary space, an advantage is that no rotating mechanism is required.

Are you tired of erecting separate rotary beams for 10, 15 and 20 meters plus dipoles or "all-band" antennas for 40 and 80 meters? If so, and if you are lucky enough to have the required area, multiple V beams may be the answer to your problem.

A simple multiband arrangement of multiple beams may be switched instantly for high gain in eight different directions (two at a time), covering 360 degrees. It may be simply built with just four radiating wires and two relays, if you can find several hundred feet of antenna space over 135 degrees on any side of your QTH.

The theory and feeding of V beams is well covered in many radio books. However, a few simplifications were employed by DL4ZC that may make your next V-beam installation easier. An unterminated V beam is bidirectional. Gain and directivity is about the same in either of two directions 180 degrees apart. If 45-degree spacing is used between legs, five radiating wires are normally required to cover all directions. In the DL4ZC installation the fifth leg was eliminated and the first leg is used, in combination with the fourth leg, to form an obtuse-angle V.

An obtuse-angle V radiates in a direction at right angles to the bisector of the obtuse angle. This combination is used to cover the missing sector A-A' in Fig. 1. Although according to theory the gain of an obtuse-angle V is not as great as with one having a proper acute angle, in actual operation the gain in directions A-A' compares favorably with gain in the other directions. The standing-wave ratio as measured in the coaxial line between the antenna coupler and the transmitter changes only a minor amount when switching to any of the available directions.

Switching of the beams is accomplished by installing two double-pole double-throw relays in a box at the apex of the beams. Standard 115-volt antenna relays were used and a plastic box of the type used to store foods in a refrigerator mounted upside down, provides a cheap all-weather housing for the relays. To keep the box dry, a low voltage is applied to the antenna relays during all periods the antennas are not used. The voltage need not be enough to operate the relays but just enough to generate a little heat in the box. A few protected small holes on the under side of the box will help prevent sweating. The wiring of the antenna relays is shown in Fig. 2. It is advisable that the relay control line be installed as far away from the antenna transmission line as possible. Different routes

> *An unterminated V beam is bidirectional. Gain and directivity is about the same in either of two directions 180 degrees apart.*

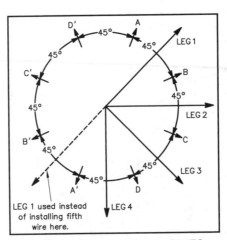

Fig 1–Directions covered by the DL4ZC V-beam array. Direction A-A' is covered by legs 1 and 4 working as an obtuse-angle V beam.

LEG 1 used instead of installing fifth wire here.

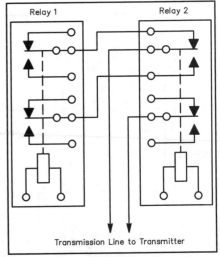

Fig 2–Circuit showing how two dpdt relays are used to switch V-beam legs. Numbers indicate connections to leg designations in Fig 1. For direction A-A' (Fig 1), both relays are unenergized as shown. For direction B-B', both relays are energized. For direction C-C', Relay 2 is energized, and for direction D-D', Relay 1 is energized.

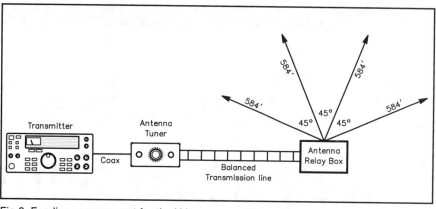

Fig 3–Feeding arrangement for the V-beam array. The antenna tuner is a conventional one for coax input and balanced output. An s.w.r. bridge may be installed in the coax line between the transmitter and antenna tuner for matching the antenna system to the coax line.

for the two lines should be followed if practicable.

The 45-degree spacing used between legs is a compromise value which permits reasonably high gain on any frequency between 3.5 and 30 MHz. The length of the four wires is not critical, so long as all four wires are exactly the same length. It is recommended, however, that the length used be approximately that shown in Fig. 3. Each leg should be approximately the same distance from ground and other obstacles. The apex of the legs and the far ends of the legs should be close to the same distances above ground.

A well-designed 600-ohm open-wire line is recommended in textbooks for use between the antenna tuner and the feed point of the V beams. Open-wire TV line should also be satisfactory for low- or medium-power transmitters. Transmitting-type Twin Lead for high power, or TV-type for medium or low power, might be satisfactory if the line is short, but

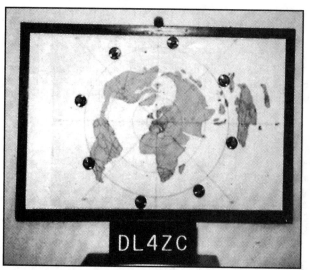

Lights may be installed around a great-circle map to indicate direction of beam in use.

standing waves will run the losses up on longer lengths. At DL4ZC the final "dressing up" of the installation consists of placing a great circle map in a frame on the wall and mounting eight light bulbs around the map spaced every 45 degrees in the primary directions of the beams. The lights are operated in pairs and are switched by the station operator at the same time the antenna relays are switched. A multiple-contact switch is used to control the operation of the antenna relays simultaneously with the lighting of lights on opposite sides of the great circle map. A quick glance at the map shows which beam is in use and what two directions it will operate in with maximum gain. Operation on all bands has netted 171 countries worked in one year. Experience has shown that on the higher frequency bands transmission reports are in general better than when using three-element rotary beams, while reception, on a relative basis, is generally poorer because of the bidirectional characteristics of the V beams.

By Ross A. Hull and C. C. Rodimon, W1SZ   From *QST*, November 1936

Plain Talk About Rhombic Antennas

The story of some experiences with haywire diamonds

Four years ago, shortly after E. Bruce announced the development of the rhombic antenna, we put up an experimental antenna of this type with the idea of working Asia. As we see it now, everything was wrong with the project except the antenna itself. We had picked the wrong time and the wrong place. Asia simply wasn't willing. There were no signals. As a result of that experience our interest in the general subject of directive antennas fell off to a mere nothing— and stayed there.

Then, in 1934 we stuck up a directive array for the 60-MHz band and found, much to our astonishment, that nice fat signals could be had with it from stations a hundred miles away at times when the signals were actually inaudible on a normal half-wave antenna. This experience gave us a big jolt because the apparent gain was out of all proportion to normal expectations. We became heavy beam-backers overnight. Ever since, we have had a pronounced leaning toward directive antennas. We have used them whenever circumstances permitted and we have looked longingly at every tree, roof, and chimney within a half mile, mulling over all the possibilities.

One big problem with any array is to decide in what direction to shoot it. This difficulty was solved recently upon hearing that Brother A. G. Hull in Sydney, Australia, had grabbed off a license and was on the air. The other big problem, to which we have never found a ready solution, is to decide just how big an array is needed to give worth-while gain. It is one thing, we have discovered, to wade through the many technical treatments of directive antennas, visualizing a great stretch of flat, swampy ground with the various wires strung up in the blue over it. Gains can be computed so readily then, and it is not at all difficult to think in terms of the R point gain per hundred feet of wire. It is a horse of a different color to stand out on the only available piece of ground–sloping, bumpy, chuck full

of trees, smeared with buildings, poles, wires and miscellaneous junk—and then to wonder what might happen to this textbook antenna under those circumstances.

Anyway, we got out the compass and a measuring tape and made a crude plan showing all the chimneys and trees of the surrounding territory. On this we superimposed models of all the antennas we could think of. Study of the layout of the many trees around the place revealed chiefly that the guys who planted them had very little knowledge of directive antennas and still

less consideration for the possible needs of future radio amateurs. The outcome, anyway, was a decision to string up a rhombic antenna of such dimensions that the transplanting of a few maples would be unnecessary. The presence of several choice 50-foot trees in the wrong places dictated that the wire would have to be threaded through two of them and wrapped around another but, we thought, that very circumstance would at least permit us to discover what does happen when such departures from the ideal are made.

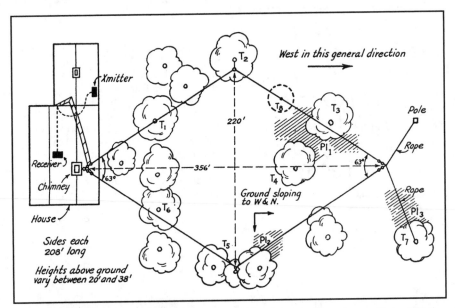

Fig 1—The layout at W1JPE showing the new 3¼-wave rhombic antenna. The original rhombic antenna discussed in the text was suspended between the chimney and cherry tree T₄. Its dimensions were exactly those used for the W1SZ rig shown in Fig 2. The antenna shown does not actually have the clean lines and symmetrical shape indicated. The wire wanders irregularly through T₁, T₂, T₃, T₅ and T₆. Also, the height varies between 20 and 35 feet. The shaded areas PI1, 2 and 3 are dense patches of poison ivy-shown to be important factors in antenna construction and adjustment. The rope between the 40-foot pole and T₇ allows small changes in the seting of the antenna. T₈ is the stump of a 40-foot tree which the authors removed by throwing a rope over it, then swaying it at its resonant frequency. The antenna works.

We shall skip now a hectic day of scrambling over slate roofs; climbing trees; threading wires through branches; getting smeared in poison ivy; unscrambling wires and ropes tangled in tree tops. These matters were important enough at the time but, like most experiences of the kind, faded into insignificance once the whole procedure *was* shown to be justified. And this particular procedure was justified. The antenna, from the very word go, functioned in a manner which we should have believed quite impossible.

The gadget we ended up with had the general shape of a diamond with sides 144 feet (approximately 2¼ wavelengths) long. The wire was about 30 feet above ground most of the way with a couple of excursions down to about 20 feet. The far end, strung up in the cherry tree T4 of Fig. 1, was terminated with several pieces of "Ohmspun" (a non-inductive resistance element manufactured by the States Company, in Hartford) totalling 700 ohms (d.c.). An ordinary 6-inch feeder with 14-gauge wire was attached to the station end of the antenna and draped over the ridge, down the wall and through the window and a couple of doors to the transmitter. A double-pole double throw relay served to switch the antenna to feeders running into another room where the receiver and operating controls are located.

First tests were made in reception—the diamond being thrown on to the receiver with a double-pole double-throw switch in place of one of the various normal receiving antennas previously used. Gains or losses were measured with the "S" meter on an HRO and all references made to R's are, therefore, in terms of divisions on the "S" meter dial. Stray pickup from the wrong antenna at the wrong time was reduced by coupling the diamond to a tuned circuit and thence to a low impedance line (the arrangement is described later) and by using a 75-ohm line from the comparison antenna. The change-over switch was therefore in a low impedance circuit in both cases.

The use of "S" meter points to express gain or loss is doubtless far from ideal but we found it preferable to the conventional business of estimating signal levels by ear or, on the other hand to actual measurement of the field around the antenna—a process made quite impractical by the existence of dense woods in almost all directions.

To get back to earth, we found immediately that signals on the line of the beam were given such a lift that, while they were painfully weak on the comparison antenna, they were extremely strong on the diamond. That, of course, is the sort of sweeping statement that we are unable to avoid. It is the sort of statement with which antenna engineers might have little patience. From the ham operating standpoint, though, it states the case. The performance of the antenna on interfering signals was similarly striking. Frequently it would be possible to hear sixth-district stations on the beam with nothing more than faint heterodyne QRM.

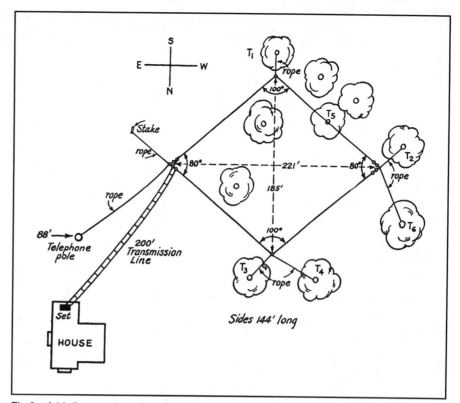

Fig 2—A bird's-eye view of the W1SZ diamond. The clear spaces on this diagram indicate dense undergrowth, brambles and a forest of second-growth trees. The antenna itself is 40 feet high at the station end and approximately 60 feet at the other points of suspension. The location of the trees used for support allows slight changes in the direction of the antenna but any change is, of course, a half-day's job. The antenna is ordinarily operated without any terminating resistor. The comparison antenna consists of two phased vertical half-waves mounted on the telephone pole.

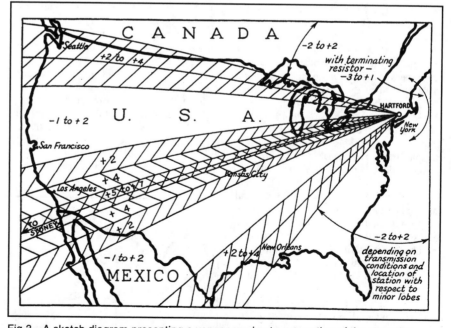

Fig 3—A sketch diagram presenting a very approximate suggestion of the reception performance of the Hayward diamond. The losses and gains indicated are R points measured on the a.v.c. meter of an HRO receiver. The comparison antenna was a conventional half-wave affair with a 75-ohm transmission line. The figures given are averages of several hundred measurements made over a period of two weeks. Though the diagram represents the performance of the antenna shown in Fig 1 it differs only in minor respects from that obtained with the W1SZ antenna.

Switching to the halfwave comparison antenna would produce, on precisely the same frequency, a fourth-district station of similar strength and with similarly inconsequential interference. The 20-meter signals from W1JPE were bumped, along the line of the beam, anything from 2 to 6 R points (estimated by the various listeners). The VK's (20-meter 'phone) over the period of a week's testing, reported us variously as the "loudest first district station," as "loud as the strongest W's from any district" and "three or four R points stronger than W1SZ."

The latter line of talk led W1SZ to throw up a similar antenna—not that it was all throwing. The location at W1SZ is even more thoroughly smeared with trees and underbrush than that at W1JPE. This circumstance, together with the fact that W1SZ decided to use copper-clad steel wire, led to many complications. A two-day struggle with the project leaves us with one firm recommendation—that if copper-clad steel wire must be strung above dense underbrush, it should be dropped into position from a blimp or other convenient type of skyhook. Threading the wire through the brush with the idea of pulling it up into position afterward is, quite definitely, the wrong idea.

To get back to cases, the W1SZ antenna also worked like a charm, bumping his signal along the line of the beam to such an extent that he now became a point or more stronger than W1JPE. The comparison antenna used at W1SZ is a pair of vertical half-waves in phase strung alongside an 85-foot telephone pole. It is an excellent antenna in the ordinary sense of the word but the new diamond ran rings around it in just as striking a fashion as did the diamond at W1JPE. The vertical antenna was, of course, preferable for work to Europe, South America, Canada, and some portions of the United States but, surprisingly enough, the diamond gave quite good general coverage in spite of the great gain along the direction of its main lobe. Using this antenna without a terminating resistor W1SZ has been able to maintain contact with VK3MR and other VK stations for 19 hours out of the 24—a performance, from this part of the world, which we have long considered an impossibility.

The point about all this rigmarole is that after reading all the idealistic technical material and after hearing vague rumors of results obtained by other amateurs we have at last had intimate experience with the rhombic antenna in ham dress. And since the experience embraces two installations under widely different conditions (both of them being similarly successful) we feel justified in trying to express our enthusiasm. Without any doubt, there are hundreds of hams with the space to put up a small diamond and the desire to pump a particularly heavy signal into some one corner of the world. Most of them would hesitate to do anything about it because they are faced, as we were, with the impossibility of dis-

covering from any of the published material whether or not the thing would be worth while. The textbooks say that a rhombic antenna with sides $3\frac{1}{4}$ wavelengths long will have a power gain of 25 over a half-wave antenna at the same height. But this leaves many questions unanswered. Over what angle, for instance, is this gain likely to be noticeable; what happens to it if the location is covered with trees; and what if the wires are actually tangled in the branches, and if the height of the wire is less than a half wave and variable along the length of the antenna—what then? What happens if the ground is irregular or sloping? And what happens to the performance if the terminating resistor is left off?

Answers to these questions, based on our own experience, go about like this: Over an angle of approximately 5 degrees the apparent power gain over a half-wave antenna in reception, particularly on DX signals, is likely to be very much more than the theoretical value—this probably resulting in cases where the vertical directivity of the antenna places the main lobe at the angle of arrival of the incoming signal. The height of the antenna above ground will influence the vertical directivity and the slight superiority of the W1SZ antenna over that at W1JPE leads us to suspect that the additional height at W1SZ has given him a lower angle of radiation in the vertical plane and, hence, a better performance on DX signals. The irregular ground and the irregular height at W1JPE has doubtless destroyed the clean form of the ideal main lobe, the effect appearing chiefly to be a slightly broader characteristic in both the horizontal and vertical planes. Trees, buildings and miscellaneous wires in the field of the antenna probably have a similar effect on the performance of the antenna, but the influence is very hard to detect. Not so long ago we should even have chopped down the family's pet trees to avoid contact between the antenna and branches or leaves. To-day we are of the impression that the matter is of precious little consequence—in the case of very long-wire antennas, at any rate.

Then there is the matter of the terminating resistor. Should facilities be available it would be possible to adjust the terminating resistor precisely and thus virtually eliminate unwanted signals from the rear of the antenna. And it would be possible,

doubtless, to improve the radiation in the forward direction by establishing and matching the characteristic impedance of the system. With our particular antennas, access to the terminating resistor is had only after a half-day's work untangling ropes and wires from the trees. A program of cut-and-try adjustment with field intensity measuring equipment is, therefore, quite impractical. We have been left with the alternatives of connecting in a 700-ohm resistor, hoping for the best, or dropping the resistor out. The chief observation is that any terminating resistor, (accidentally we have tried 300, 500 and 800 ohms) simplifies feeding the antenna since, under those circumstances the system will take plenty of power without tuning the feeder. Elimination of the termination resistor makes it necessary to tune the feeders but the performance in the forward direction is quite similar. The terminating resistor, even if incorrectly adjusted, gives a drop of several R points to signals arriving from the rear of the antenna. The reduction in noise coming from the rear is also noticeable.

Our most recent experience with this type of antenna has been in the erection of a larger system ($3\frac{1}{4}$ wavelengths on a side) at W1JPE in the attempt to blot out the W1SZ signal in Australia. The new antenna, though larger, is considerably more irregular in its various dimensions than the first version and probably because of that its performance is not quite what we had expected. The main lobe and the two first secondary lobes give us a performance in reception similar to that shown in Fig. 3. This chart, indeed, is the result of several hundred readings taken on the HRO "S" Meter while comparing the $3\frac{1}{4}$ wavelength diamond against a half-wave comparison antenna. It differs from the characteristic had with the $2\frac{1}{4}$ wavelength antenna only in the distribution of the minor lobes. It represents, in short, just about what one might expect from a very haywire diamond between 200 and 300 feet from tip to tip.

And so, after all these very general statements, we reach the point where we can suggest with all the emphasis we can command that any ham who has a hankering to pump big signals in one or two particular directions, and who has any chance at all to borrow or rent the space, is doing the wise

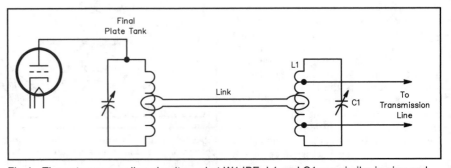

Fig 4—The antenna coupling circuit used at W1JPE. L1 and C1 are similar in size and rating to the coil and condenser used in the plate circuit of the final amplifier. The coupling of the two-turn link and the setting of the taps on L1 are varied until the desired load is obtained.

thing if he cancels his order for two half kilowatt bottles and puts up a diamond instead. Don't mind the trees and the underbrush, don't mind the buildings and the clothes line—just string the thing up and shoot. Remember though, that it is quite ridiculous to use such an antenna for transmission while using a piece of wire around the picture rail for receiving. It is utterly impossible to exploit the possibilities of the antenna without a change-over switch or relay which will permit using the antenna for reception. The method of coupling the antenna to the receiver is also important.

We suggest setting up a tuned circuit consisting of a 35-mmfd. midget variahle condenser and an 8-turn coil of bare wire one inch in diameter, coupling this with a 2-turn link to the terminals of the receiver. The feeders from the antenna should then be clipped across about the middle four turns of the coil. A somewhat similar arrangement, shown in Fig. 4, is suggested for the transmitter.

Possibly the most important feature of all is that the rhombic antenna operates effectively over a very wide frequency range. It is the one type of directive antenna that functions without the need of any adjustment or change on, say, the 40-, 20-, and 10-meter bands. Further, as the frequency is increased the vertical angle of radiation is decreased. Result—a hot performance on three bands. In practice the WlJPE antenna is an absolute whizz on 28 MHz, even giving greater gains than those had on 14 MHz. Time and again we have had thoroughly satisfactory 'phone contacts with stations along the line of the beam at times when the signals simply did not exist on the normal antenna. It is all rather hard to believe.

From *QST*, August 1967

That W1AW Rhombic

Back in 1937, very shortly after the League purchased its 7-acre plot on Main St. in Newington, for the erection of a headquarters station, the first thing accomplished was the erection of a rhombic ("diamond") directional antenna, aimed at the west coast. Five (four of them 60' tall) western red cedar poles were purchased in Oregon and transported here at some expense, and planted at points on the property laid out in advance by a surveyor (W1OKY) to orient all projected W1AW antennas in accordance with a preconceived plan. The rhombic was the first antenna to go up. It consisted of 1400' of wire, not counting the feedline, and was left unterminated.

The rhombic really worked. A visitor to W1AW before the building was complete hooked his portable 7-watt rig to it, worked Australia. In the fall of 1938, when W1AW went into operation officially from its new location, fantastic signal reports were received from the mid-Rockies, from the west coast, and from the South Pacific. During the intervening years until headquarters itself moved to 225 Main St. from LaSalle Road in West Hartford, the problem of west coast coverage for W1AW was never serious. Whenever someone complained they couldn't hear the station, we could usually demonstrate that W1AW was laying down a wicked signal on 20 meters.

But the new headquarters building spelled the end of the rhombic—temporarily, at least. A three-element beam was erected atop a 60-foot tower, and tests seemed to indicate that it was the equal, or nearly the equal, of the rhombic and had the advantage of being rotatable. As time went on, however, listeners on the Pacific coast began to complain that WlAW was not copiable and it was eventually decided to reerect the rhombic.

The five original poles were still up, and experts from the power company pronounced all but one of them in solid condition, after 30 years. One pole had been moved slightly so as not to be in the way of building construction, but not so much so as to affect the rhombic orientation materially. The condemned pole was replaced by a steel tower.

By the time you read this, the rhombic should he back in operation. It will be used primarily on 20 meters and should enhance W1AW's signal strength on most of the west coast.

Fig 2—Early photo of the W1AW rhombic. As a result of vandalism, the far-end support fell during Memorial Day weekend in 1987. At that time the antenna was only used as a backup for the monoband Yagis that were in daily use. In the summer of 1989, the remaining supports and the last of the rhombic wires were taken down.

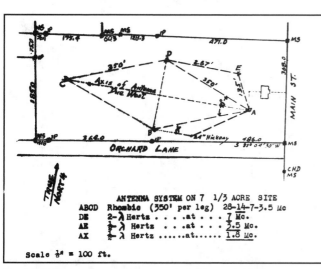

Fig 1—Diagram of the W1AW Rhombic that was erected at 225 Main Street in 1937. Orchard Lane has been renamed as Starr Avenue.

The Tilted Half-Rhombic Antenna

AA2PE analyzes a unique multi-band wire antenna with switchable pattern.

This antenna is designed for general coverage use on the HF bands from 30 through 10 meters. It is broadband so that matching is possible without the need for tuned matching networks. For use at AA2PE, the antenna had to fit in a restricted area of about 100 × 100 feet. It can be built with materials readily available to amateurs. The most expensive part of the antenna are the two broadband matching transformers.

Description

The tilted half-rhombic antenna resembles the more familiar "sloping V" antenna. The difference is that the tilted half rhombic is fed at an end, and a sloping V is fed at the center (apex). The feedpoint and terminating point of the antenna are placed on diagonally opposite corners of the given area. To keep the supporting structure from being an obstruction, it is located in one of the remaining corners. An illustration of the tilted half-rhombic antenna is shown in Fig 1.

The antenna, as installed at AA2PE, has 27.2 meters of wire in each leg. At 30 meters, this provides a leg length of nearly 1 l. At 10 meters, the leg length is slightly over 2.5 l, as calculated from the equation:

$$\text{Leg length} = 150 \, (N - 0.05) \, / \, F_{MHz}$$

where N is the number of half wave-

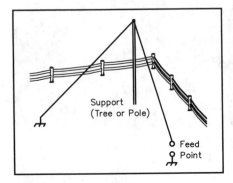

Fig 1—Layout of AA2PE tilted half-rhombic antenna.

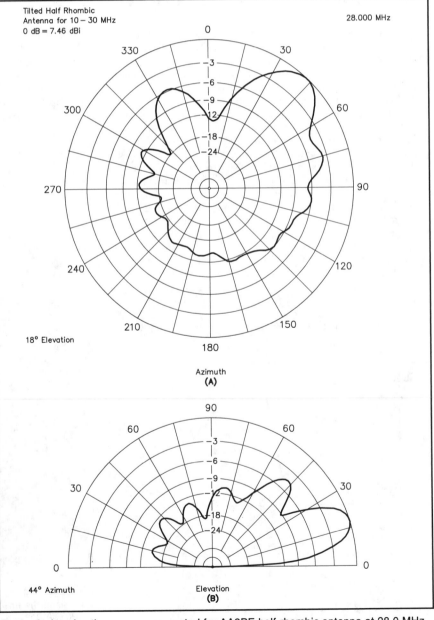

Fig 2—At A, azimuth response computed for AA2PE half-rhombic antenna at 28.0 MHz. Maximum radiation is along axis drawn from feedpoint to terminating end. At B, elevation pattern at azimuth of maximum radiation.

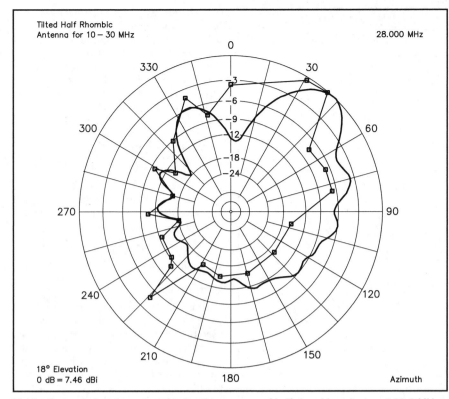

Fig 3—Computed and measured azimuth response of half-rhombic antenna at 28.0 MHz.

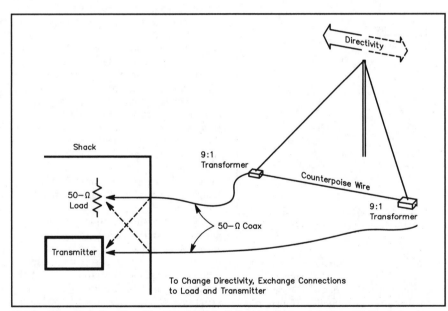

Fig 4—Changing directionality of AA2PE half-rhombic by exchanging 50-Ω coaxes in shack.

Table 1
Computed Feed-Point Impedance and Computed Versus Measured SWR of AA2PE Tilted Half-Rhombic Antenna

Freq (MHz)	Series R	Series X	Calculated SWR	Measured SWR
10.1	484	–132	1.34	1.10
14	477	–222	1.60	1.05
18.1	423	–96	1.26	*
21	396	–152	1.46	1.25
24.9	364	–115	1.72	*
28	268	–148	1.94	1.40

* Data unavailable

lengths in each leg.

The height of the apex in the AA2PE antenna is approximately 9½ meters. With 27.2 meters of wire in each leg, the included angle between the wires is approaching a more acceptable value for radiation lobe alignment from each leg. More wire may be used, and is encouraged, especially to make each leg longer than 1 l at the lowest desired operating frequency. If the increase in wire length is significant over that given here, the height of the apex must be increased to get a quality pattern. The approximate angle at the apex between the wires should be maintained at about 90° to 100°. More information on radiation tilt angles and lobe alignment in V and rhombic antennas may be found in Chapter 13 of the 15th Edition of The ARRL Antenna Book.[1] This antenna may be considered as two long wires end to end, and at an angle from each other. Radiation from a long wire is conical, with the wire as the axis.[2]

Pattern

The antenna's major lobe radiation is along the axis drawn from the feedpoint to the terminating point, and in the direction of the terminating end (See Fig 2). A computer simulation plot of the antenna has been made using the AO antenna analysis and optimization program.[3] [Note that with wires close to the ground, AO, a variation of MININEC, will not give accurate results, typically inflating gain by as much as several dB. Running NEC for the same antenna verifies this.—Ed.]

Field-strength measurements were taken at a 2 km radius around the proto-type antenna at 28 MHz. To get a comparison between model results and actual results. the data was entered on the plot created by AO. See Fig 3. The receiving antenna was a vertical whip mounted on an automobile and the measurements were taken with a calibrated field-strength meter. The field strength meter readings in microvolts were normalized and converted to dB for plotting.

As Fig 3 shows, the actual radiation pattern is very well correlated to the model results. There are a few minor departures from predicted values. There is a spurious lobe appearing at 225°, exactly in line with the radiation axis. This represents radiation in the opposite direction due to a reflected wave on the antenna. The reflected wave is a result of imperfect match between the antenna characteristic impedance and the matching transformers and their loads. The reduced field strength off the "east" of the antenna is believed to be due to measurement error, caused by being in an area of higher vegetation where the measurement was taken.

There is a small, unaccounted for, shift in the main lobe's position. Power lines and metal rain gutters in the vicinity reradiate the energy from the antenna, causing a distorted pattern. As with other sources of measurement error, the actual antenna pattern will not always be identical to the model results. Hopefully, the differences

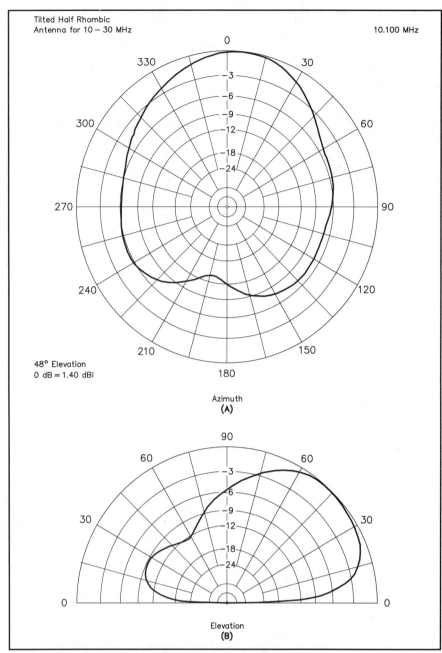

Tilted Half Rhombic
Antenna for 10 – 30 MHz

10.100 MHz

48° Elevation
0 dB = 1.40 dBi

Azimuth
(A)

Elevation
(B)

Fig 5—At A, azimuth response for AA2PE antenna at 10.1 MHz. At B, elevation response.

Impedance Matching

Broadband operation is accomplished by terminating the end of the antenna into its characteristic impedance. The tilted halfrhombic has an impedance about half that of a standard rhombic, around 400 Ω. To simplify the matching network design, a terminating impedance of 450 Ω has been chosen. The feed line is connected to the antenna through a 9:1 unbalanced to unbalanced transformer. The terminating resistance is provided by a standard 50-Ω dummy load, through a 50-Ω coaxial cable and a 9:1 transformer. Ground return is provided with a counterpoise wire along or just a few inches below ground. No other radials or additional ground points are required,

beyond what should already be established for safety reasons. The broadband transmission line transformers used were of the Guanella design presented in Transmission Line Transformers.[4] This critical part of the antenna must be carefully built for optimum performance.

The input impedance was measured with an RX meter connected directly at the feedpoint, without the matching transformers. Since the feedpoint is on the ground, this is very simple to do. The measurements are given in Table 1. A noise bridge could be used to make impedance measurements, if one has a battery operated receiver. It would be more convenient to make the measurements through the matching transformer to determine the departure from a 50-Ω match. It may also be possible to determine matching network loss by observing the depth of null, compared to a pure 50-Ω load.

Depending on how much feed line you have, the measured SWR in the shack may be significantly lower. For instance, I measured an SWR of 1.4:1 on 10 meters, where the impedance measurement indicated that the SWR should have been 1.94:1. The difference is due to attenuation in the feed line of the reflected wave as it travels back to the transmitter.

Since the antenna is directive, it's desirable to have some control over the radiation direction. A coworker suggested bringing the 50-Ω coaxial feed line into the shack from both ends of the antenna, through the 9:1 transformers. If access is provided to the two feed lines, the connections can be exchanged between the dummy load and rig. This turns the antenna main lobe over by 180°. A diagram of this is shown in Fig 4.

Performance

Does the tilted half-rhombic antenna work? Yes, it does! On the air, it has been used for DX and stateside QRP contacts on 10 through 30 meters. These contacts were made during a few weekend hours of operation in the Fall of 1993 and with an output power of only 5 W or less. With such low power, the received RST reports were not astounding, but the signal was copied successfully. Stations as far away as 15,000 km have been contacted as well.

I recognize that this antenna is not the perfect antenna. My goal was to provide a structure that could be used within the limitations of space and placement, and to get on the air with decent performance. Does the antenna have its operational limitations? Yes, it does! Attempts were made unsuccessfully several evenings to contact a station in Antarctica on 30 meters. On this band, though, the antenna has reached its limit of performance. As the frequency of operation is lowered, the performance of the tilted half-rhombic drops off. This is illustrated in a plot of the antenna at 30 meters. See Fig 5.

Comparing the radiation pattern of this antenna to a terminated sloping V demonstrates a significant gain and pattern quality advantage, resulting from the feedpoint being taken at the end rather than at the center (or apex). The antenna analysis program AO was again used to make this comparison. The same antenna structure is modeled, with the feedpoint at the center, and then with the termination at either end. The results are shown in Fig 6. It can be seen that the gain and pattern quality of the sloping V is considerably worse than the tilted half rhombic.

Receiving tests were conducted on 13 meters by establishing a listening schedule of selected short-wave broadcast stations within the main radiation lobe of the pattern. A half-wave dipole was installed, and A/B comparison tests were conducted. A measurement of receiver AGC voltage with a digital voltmeter was made up to twenty times for each antenna, 10 readings for A, 10 for B, 10 for A, and so on. The data collected

are slight enough so that one can rely on the model results to predict actual antenna performance.

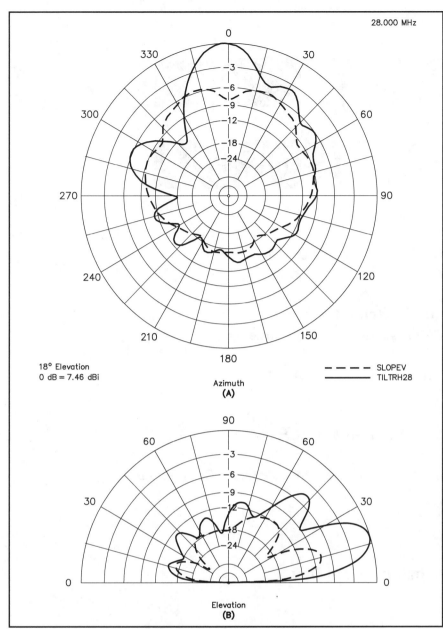

28.000 MHz

18° Elevation
0 dB = 7.46 dBi

Azimuth
(A)

- - - - - SLOPEV
———— TILTRH28

Elevation
(B)

Fig 6—At A, comparison of half-rhombic (solid line) and terminated sloping-V antenna (dashed line) at 28.0 MHz. Note that pattern for half-rhombic has been rotated to line up at 0° azimuth. At B, comparison of elevation patterns for same two antennas.

Table 2
Gain as a Function of Included Angle for AA2PE Tilted Half-Rhombic

Freq (MHz)	AA2PE Gain (dBi) (95° incl angle)	AO Incl Angle (°)	AO Gain (dBi)
28	7.5	106	8.1
24.9	7.1	102	7.5
21.0	7.0	97	7.0
18.1	6.0	88	6.0
14.0	4.0	81	4.7
10.1	1.7	59	2.5

show measured gain of as little as 2 dB and as great as 11 dB, with the average about 5 dB. Many factors contribute to the variability in the data—mostly the arrival angle and polarization of the received wave. These factors are variable over time and station.

Pattern and Gain Optimization

The angle between the two legs of the antenna, called the included angle, plays an important role in the antenna's performance. Using this knowledge, some may have the flexibility to place the antenna in a more favorable position. As it was placed at AA2PE, the included angle is approximately 95°. One could predict, from the information in Reference 1, that the optimum included angle should be 110° for a leg length of 2.5 λ. This prediction was tested with the antenna modeling program *AO*, by invoking its antenna optimization feature. Left to its own problem solving, *AO* gave a result of 106°, with a gain increase of 0.6 dB on 10 meters. For each frequency, there will be a different optimum included angle. A comparison of the AA2PE antenna and the *AO* antenna optimized at each frequency is shown in Table 2.

Conclusion

By moving the feedpoint of a terminated sloping-V antenna from the apex to an end, a considerable improvement in antenna performance can be achieved. This antenna provides a good impedance match across the HF spectrum without the use of a Transmatch. The SWR is relatively low on the feed line. The tilted half-rhombic is a very good candidate for a Field Day antenna, since the counterpoise need not be buried, only one support is required, and the feedpoint is at ground level. No ground rods are necessary for the antenna to function properly. The termination may be removed, yielding a more omnidirectional pattern. This may be desirable for Field Day use.[5] If more space is available and a higher apex can be provided, the antenna's operating range may be increased to cover the entire HF spectrum and to improve gain. Remember, the transmission line transformers have to be made aufficiently broadband as well.

Acknowledgments

The author wishes to give special recognition and thanks to Palemon "Dubie" Dubowicz for his suggestion to bring both ends of the antenna into the shack to allow changing the pattern direction. Dubie also provided much assistance in the area of broadband transformers, theoretical and practical, and supplied some units used for the initial prototype. The original vertical halfrhombic design on which this antenna is based was developed by Dubie for the US Army Signal Corps. The Army antenna kit (AS-303) can be erected within 20 minutes by two soldiers, and is truly a field antenna. I hope this effort will return to Dube a rekindled desire to obtain his Amateur Radio license, which he long ago let lapse.

Notes

[1] G. Hall, K1TD, Editor, *The ARRL Antenna Book* (Newington: ARRL, 1988).
[2] J. D. Kraus, *Antennas*, Second Edition (New York: McGraw-Hill, Inc, 1988), p 502.
[3] *AO* 6.0 Antenna Optimizer software, Brian Beezley, K6STI.
[4] Sevick, Jerry, W2FMI, *Transmission Line Transformers*, 2nd Edition (Newington: ARRL, 1990)
[5] 1t is recognized that the current that would have been dissipated by the terminating load is now reflected, and will raise the SWR. Experiments at AA2PE indicate that if the wire is sufficiently long (greater than 1 l per leg) the SWR remains below 2:1. The caution is that this measurement depends on the length of the feed line and its loss, and whether the load is merely disconnected from the terminating feed line, or if the feed line itself is removed. The unterminated antenna should be tested for a specific installation before counting on it to provide an acceptable SWR.

CHAPTER SIX

WIRE BEAMS

By Jarda Dvoracek, OK1ATP   **From *QST*, October 1975**

160-Meter DX With A Two-Element Beam

It takes more than desire and imagination to be an effective DXer on "top band," and the impressive record of OK1ATP is proof that some hard work and dedication went into his 60-country total on 160 meters. No doubt his success results in part from the antenna system used at OK1ATP—a two-element fixed-heading wire beam, the details of which are given in this article.

Many antennas for DXing on 160 meters have been tried at OK1ATP, but the best results came when using the antenna described here—a two-element inverted-V beam of the driven variety. It is fed with balanced TV line, as shown in Fig. 1. The transmitter is matched to the line by means of a simple tuned network. The beam

> *There is a significant front-to-back ratio with the wire beam.*

is oriented to provide maximum radiation to the west. A separate antenna — a single-element inverted V — is used for coverage to the east.

My signals in England are two S units better with the beam than with the single inverted V. Signal reports from the U.S. show the beam to be one S unit superior to

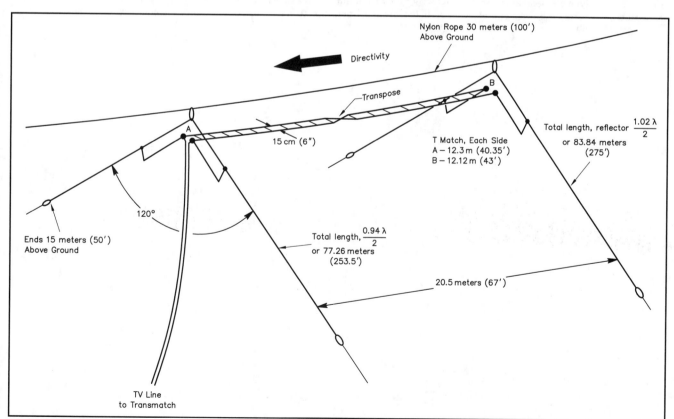

Fig 1—Principal details of the OK1ATP 160-meter two-element beam. Elements are suspended from a heavy nylon rope in which knots were tied to maintain the element spacing. Both are driven by means of the transposed balanced line, as shown. The main run to the station is open TV line. Dimensions are for operation near 1825 kHz.

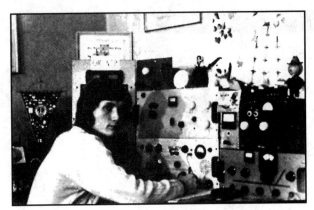

Completely home built station, OK1ATP. Jarda is known far and wide on 160 meters for his outstanding signal.

openings and be present when activity takes place. Because the band is such a challenging one, my interest remains high. On the average, I devote 20 nights a month to operating — from 2000 to 0100 GMT, and also around sunrise time here. Most of my transmitting is done between 1825 and 1827 kHz in the "DX window."

It is reasonable to conclude from the results I have obtained that the two-element beam is performing well.

My signals in England are two S units better with the beam than with the single inverted V.

the inverted V. My inverted V is the better antenna when working east . . . approximately two S units better than from the backside of the beam. This indicates that there is a significant front-to-back ratio with the wire beam.

Nylon rope is used to support the beam. The rope is stretched tight between two supports, and the driven element and reflector droop downward from the rope. Knots are made in the rope at appropriate points to hold the elements in place and to keep the T sections the desired distance from the elements (Fig. 1). Open-wire phasing line is used to connect the driven element to the reflector. It is transposed as shown in the drawing. Homemade plastic spacers are used to separate the wire in the T sections, phasing line, and transposition point. The European name for the rope and spacers is "Silon."

Results With the Beam

I have been trying to obtain the DXCC award on 160 meters since 1968 and thus far have 60 countries on my "worked" list. All of the station equipment is homemade, as seen in the photograph which was taken in 1972. Results have been good since erecting the beam, but it takes many hours of listening and operating to catch band

I have worked 207 U.S. stations, and all call areas other than W7, KL7, and KH6. I have worked VO1, VE1, VE3 and VE7 stations. With W1HGT I've had 60 QSOs so far, and with W1BB there have been 54. I have heard a considerable number of countries which I have not worked.

It is reasonable to conclude from the results I have obtained that the two-element beam is performing well, even though it blew down during our hurricane of December, 1974 when the nylon support rope broke! It has since been rebuilt and is working nicely. It is my hope that this information will be of interest to other 160-meter operators around the world.

Beams With Inverted-V Elements

It might be of some interest to note that inverted-V type elements can be used as a full-size beam on an average lot. See Fig. 1. This arrangement is in use at my QTH and gives a good account of itself. My regular full-size doublet for 40 meters is utilized as a boom for the inverted-V beam. This, of course, permits full quarter-wave spacing for the three-element beam. The insulators are made from epoxy-board material with the copper etched off. A common cable clamp at the center of the regular dipole is used to support the cable and the inverted-V dipole via a 6-inch length of nylon rope. Both insulators are made the same way. The short length of rope allows the lower insulator to be at right angles to the top insulator. Most hams have at least one doublet in the air. With this arrangement, they can get a beam at a minimum of cost. It works, it's cheap, it's easy to construct; in fact, this type of beam could be supported by a peaked roof top, making an ideal concealed antenna. — *Tom Marshall, W5LT*

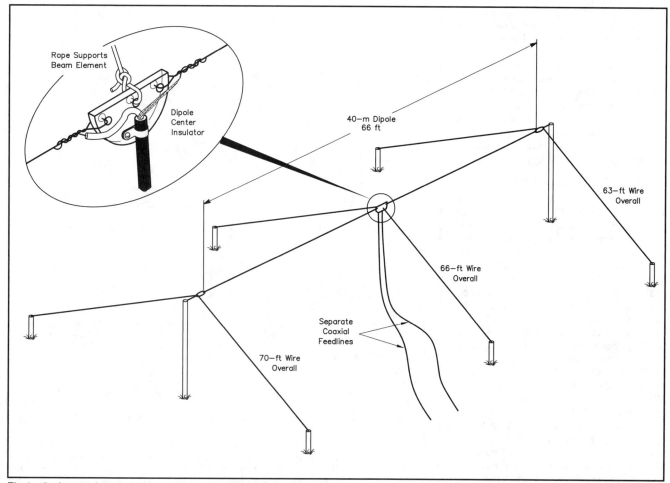

Fig 1—3-element inverted-V beam supported by ordinary dipole antenna.

By Edward Peter Swynar, VE3CUI   From *QST*, May 1984

40 Meters with a Phased Delta Loop

Large 40-meter Yagis can set you back a week's salary and take a month to install. A bidirectional, 2-element Delta Loop array provides a better way to snare some DX at modest cost.

A station in the U.K. was heard to say, "Forty DX separates the men from the boys." In line with this, it is fortunate for the home-construction crowd that 7 MHz is an area where the mind must often rule over matter! In pursuit of 40 meter DX, some amateurs have embraced the costly "consumer approach." Others have resigned themselves to the likes of the simple and relatively ineffective inverted-V antenna—coupled to the omnipresent kilowatt amplifier. There is a better way!

With moderate property dimensions, some trees (perhaps), wire, coaxial cable and a bit of patience, it is possible to build an excellent gain type of array. It can be switched to either of two directions. It is inexpensive and effective for working long-haul DX. I will refer to it as the "2-element, 90-degree-phased Delta Loop."

The Case for Phased Loops

Literature abounds regarding the cardiodal pattern of 2-element 90-degree phased vertical antennas with 0.25-wavelength spacing. A gain of 3 dB is available over a single 0.25-wavelength vertical element.[1] But, since such an element has a minus gain of 1.8 dB over a dipole, one can realize a 1.2-dB gain over a dipole when using two verticals that are phased. The major advantage of the vertical 2-element array is, therefore, the low radiation angle and the directivity (at the expense of many buried copper radials).

With 90°-phased dipoles there is, relatively speaking, more gain and less wire. Again, each dipole element by itself has no gain (using dBd as a reference). Also, this type of array must be fairly high above ground to be an effective DX antenna on 40 meters. Now, consider the phased Delta Loop arrangement of Fig. 1. By virtue of the feed points on each element, the array is vertically polarized and produces a low

angle of radiation, as with the phased vertical system. Furthermore, each loop (by itself) offers a 2-dB gain over a dipole (3.8 dB over a single 0.25-wavelength vertical). Imagine the benefits of two such gain-style loops, positioned properly and driven in

combination to enhance the already existing gain of a single loop element.

Construction

Your specific situation will dictate the precise shape of your loop. Nevertheless,

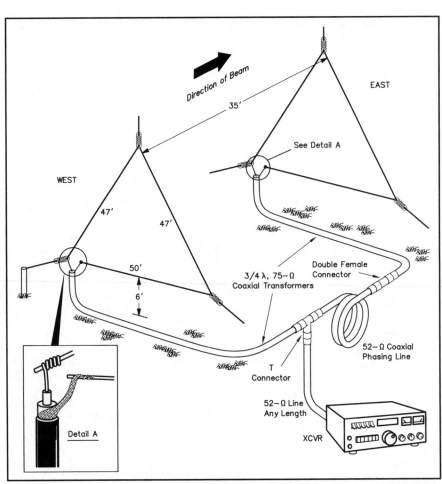

Fig 1—Illustration of the final arrangement chosen at VE3CUI for the phased 2-element Delta Loop array. Corner feed, apex up (as shown) yields vertical polarization and a low radiation angle.

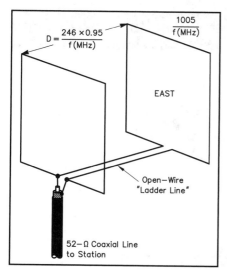

Fig 2—Arrangement for the unidirectional loop array that was used first at VE3CUI.

the length of the wire for each element should be taken from the standard loop equation—L(ft) = 1005/f(MHz).[2] I like to add approximately 2 feet of additional wire to facilitate final adjustment for lowest SWR. The element spacing (based on freespace conditions) is obtained from L(ft) = 246/f(MHz).

I first used the feed method seen in Fig. 2. This system has the advantage that use of costly coaxial line is restricted to a single run of 52-ohm cable from the antenna to the ham shack. Also, the balanced phasing line helps to preserve the symmetry of the array. I'm sure this could be improved further by inserting a 1:1 balun transformer at the feed point. The disadvantage of this method is seen when trying to reverse the directivity of the antenna: I must go outside the shack, remove the coaxial feeder from one loop and connect it to the other loop. This is no fun whatsoever when the band is open to two directions at once during a cold January morning!

My present feed system is that of Fig. 1. It is an odd-multiple expansion of the conventional 1/4-wave matching transformer, the type used for matching to single loops that are fed with 50-ohm line. I tripled the length of the 75-ohm line section to 3/4 wavelength. This was a convenience because of the distance between the shack and the most distant loop. Two equal lengths of 75-ohm coaxial cable are used as transformers (one per loop). The line length is determined by

$$L \text{ (feet)} = 0.66 \left[\frac{246}{f \text{ (MHz)}} \right] \times 3 \qquad \text{(Eq 1)}$$

when the coaxial cable has solid dielectric rather than foam material. In this case, the velocity factor of the line is 0.66. This factor will be different if you do not use solid-dielectric polyethylene line.

Adjustment

The loops should be adjusted separately

for resonance. Attach a 3/4-wavelength transformer to one loop, then connect the free end of the transformer to a random length of 52-ohm line (through an SWR indicator). Attach the remaining end of the 52-ohm cable to your transmitter. While using the least power possible to obtain an SWR meter indication, adjust the loop length for a 1:1 SWR. [*Safety first! Do not touch a "hot" antenna. Take the rig to the antenna site, or have a friend switch it on and off for you during the tests.—Ed.*]

On completion of this procedure, repeat it with the remaining loop. I do not recommend that you "stagger-tune" the loops in the hope of obtaining increased bandwidth; one loop should be the electrical twin of the other one. I have found, also, that both loops should be the same shape and height above ground, and as perfectly spaced apart as possible. This suggestion may seem extreme, but best results will be had later on if some pains are taken during installation and adjustment.

With the loops installed in their final positions, it is time to add the 52-ohm coaxial phasing section. The length is determined by Eq. 1, but do not multiply by 3, as in the equation, since the line will be an electrical quarter wavelength rather than 0.75 wavelength. This phasing line can be rolled up and taped so that it won't occupy a lot of space in the ham station. This phasing line should be placed in series with the feeder that connects to the loop element that will serve as the forward radiator, since it will be the element that will require the 90° lag. The remaining end of the phasing section is connected (by means of a coaxial T connector) to the end of the feeder that goes to the other loop element. The third port of the T connector is used to mate the feed system to the transmitter and receiver via a short run of 52-ohm coaxial cable. Switching of the directivity is done manually in the shack by transposing the ends of the T connector that go to the feed system. Faster switching can be had by using a coaxial relay or manual switching method. For my needs, it was easy to grow used to reversing the two PL-259 plugs by hand.

The layout of my 50 × 80-foot lot is such that the directivity of the array is NNE or SSW. This has been good for DX from Europe and the South Pacific. One loop is held aloft by means of a tall tree. The other loop is supported by my 48-foot tower, and it is spaced 10 feet away from the tower.

The SWR curves differ between "beaming east" and "beaming west." See Fig. 3. I feel that the problem is caused by the aluminum siding on the house, which is close to one of the loops. Despite this annoyance, the system is flat across the part of the band that interests me—the DX segment.

Results

The bulk of my DXing is done at a power level of 500-W dc input. The exception was my first QSO with the antenna, during which I was using 50 W: I received an RST 559 report from 3B8CF on Mauritius Is-

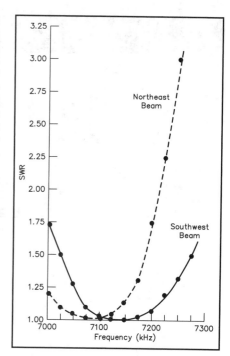

Fig 3—SWR curves developed at VE3CUI for the 2-element loop array. Note that the curves show a disparity. This may be the result of one loop being in proximity to the aluminum-sided house.

land, despite the pileup bedlam.

The front-to-back ratio of the array appears to be roughly three S units (18 dB) over long DX paths. Over short paths (interstate or interprovince), do not expect much by way of F/B ratio. Close-in contacts will be more satisfactory with a high angle radiator, such as a single loop from this array, or a dipole, can provide.

The phased loops certainly "hear" the signals better than other antennas I have used. Also, I seem to receive longer band openings than with other types of antennas I have used. I have been gratified a number of times by comments such as, "Best signal from North America, OM." I have not made performance comparisons against a reference dipole, but I received the substantially stronger signal report during a four-way DX QSO that included two local hams. One was using an inverted V, and the other had a single Delta Loop. One fellow had a report of "inaudible" (inverted V at 40 feet), and the ham with the Delta Loop was barely discernible in the noise. My report was 5 × 8.

Conclusion

Despite the low antenna height and cramped space, the 2-element phased loop array is a superlative budget-saving performer. I hope some of you will investigate the DX potential of this simple antenna. Certainly, you will experience the same kinds of pleasures I have while chasing DX on 40 meters!

Notes

[1] Gain figures are unproven and are theoretical.
[2] m = ft × 0.3048.

By Al Christman, KB8I, Tim Duffy, K3LR and
Jim Breakall, WA3FET

From *QST*, August 1994

The 160-Meter Sloper System at K3LR

Sure, doesn't everyone have a 190-foot tower in his backyard? With help from his friends, K3LR made his 160-meter dream come true.

During the spring of 1992, we began the development work on a 160-meter antenna system using half-wave slopers. This antenna was to be installed on the 190-foot tower belonging to Tim Duffy, K3LR. Our goal was to build an array that would provide forward gain in any one of several switch-selectable compass directions. Good rejection of high-angle signals was an important requirement, as was overall efficiency. We describe in this article a quick review of the theoretical design process, and then discuss the construction, testing and operation of the actual array.

Background

Perhaps the best-known directional antenna using sloping dipoles is that of Dave Pietraszewski, K1WA.[1] His design utilizes five identical λ/2 slopers spaced uniformly around a mast tall enough so the dipoles descend toward the ground at an angle of 60° below horizontal. All five radiators are fed with equal lengths (slightly over 135°) of coax. Only one element is driven at a time, and the other four open-circuited transmission lines function as loading inductors so that all of the passive elements act as reflectors.

A second type of directional array using slopers was described by Dennis Mitchell, K8UR.[2] This antenna system requires only four dipoles, and the lower half of each radiator is "pulled in" (see Figure 1) so that it slopes back toward the base of the tower. Changing the geometry this way produces a signal whose polarization is almost entirely vertical, depending on the exact disposition of the wires. All of the elements in the K8UR antenna are driven with equal-amplitude currents phase-shifted by either 0°, 90°, or 180°, just like the classic "4-Square" phased-vertical array. As with all such phased arrays, the challenge is to get the feed currents into each element exactly right, so the full potential of the array can be realized, particularly for front-to-back ratio.

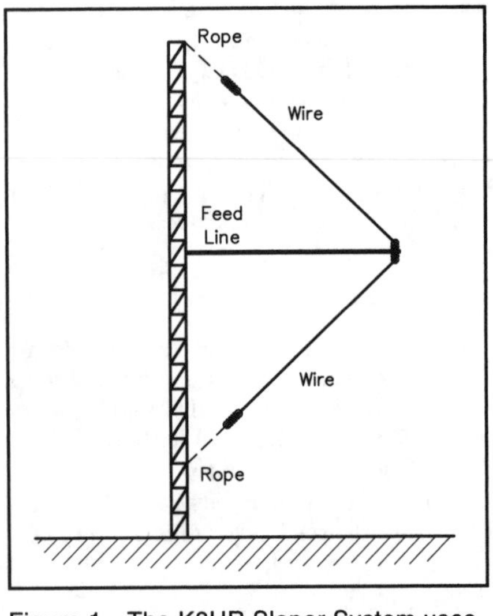

Figure 1—The K8UR Sloper System uses four identical half-wave sloping dipoles spaced uniformly around a tall mast. The lower half of each dipole is pulled in toward the tower.

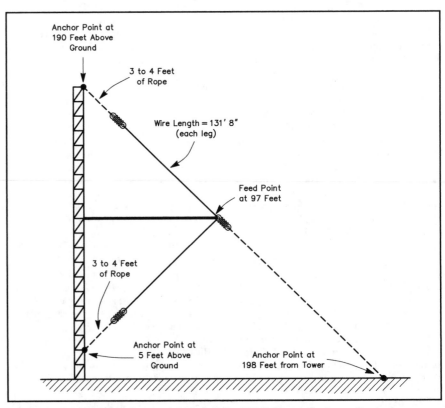

Figure 2—Detail of one element (of four) of the K3LR Sloper System, after tuning adjustments to compensate for insulation on element wires.

Design

The initial design of the K3LR antenna combines some of the best features from both the K1WA and K8UR arrays—the mechanical simplicity of the K8UR design, with the straightforward feed system of the parasitic K1WA array. The physical appearance of the K3LR system is close to that of K8UR's, consisting of four identical bent dipoles spaced at 90° intervals around a tall mast. Figure 2 shows the layout of a single element, including all of the final dimensions. This array is electrically similar to that of K1WA, because only one element at a time is driven. The remaining three dipoles are inductively loaded by open-circuiting the far end of each individual feeder, so that all three act as parasitic reflectors. Only four elements are used (rather than five as described by K1WA) because modeling with *ELNEC*[3] indicated that there were no performance advantages to be gained by using the extra dipole. Figure 3 illustrates the principal-plane radiation patterns of this early-stage K3LR antenna system.

For the 160-meter array, experimentation with *ELNEC* indicated that the best combination of gain and front-to-back ratio would occur when the parasitic elements were loaded with about 100 Ω of inductive reactance. To produce this amount of loading, an electrical length of 153.45° of open-circuited *lossless* 50-Ω line was needed at the center of each element. Because of the long cable lengths, we chose to use RG-8X (instead of RG-213) for the feeders in order to reduce the suspended weight. With a velocity factor of 78%, the calculated physical length of each transmission line is 177.74 feet at the design frequency of 1.840 MHz.

Since the four RG-8X feeders actually have a small amount of loss, even at

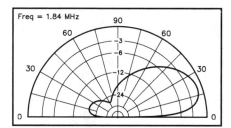

Figure 3—K3LR Sloper System, early stage modeling (elevation plot). The outer ring is 6.0 dBi, maximum gain of array is 5.06 dBi.

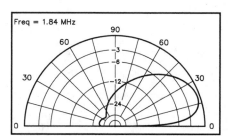

Figure 4—K3LR Sloper System, including feed-line losses in computations (elevation plot). The outer ring is 5.0 dBi, maximum gain of array is 4.57 dBi.

160 meters, the impedance at the center of each reflector due to the open-circuited lines was not purely inductive but had become complex, with a value of 25.76 + j93.45 Ω. When this corrected loading impedance was substituted into *ELNEC*, the radiation patterns that resulted were rather surprising. Figure 4 shows that there is a small reduction in forward gain—but a dramatic improvement in front-to-back ratio!

We wanted even more gain, so a decision was made to add four elevated quarter-wave radials to the antenna system. These radials would be horizontal, mounted about 10 to 15 feet above the ground, and all four were to have their inner ends connected directly to the tower. Since the radials were close to the ground in terms of wavelength, we decided to use *NEC*[4] with its Sommerfeld/Norton ground model (rather than *ELNEC*, with its simplified Fresnel reflection coefficient ground model) for the remainder of our computer analysis. In order to be really precise, we felt that it would also be necessary to include all of the small antennas (relatively speaking, of course) which were already mounted on the K3LR 190-foot tower. So, two full-size 3-element 40-meter beams (at 100 and 190 feet), a 6-element 10-meter beam (at 198 feet), and a 3-element 20-meter beam (at 60 feet) were added to the *NEC* computer model, along with the tower, the four slopers and the four elevated radials.

Figure 5 is a drawing of the *NEC* model for the complete tower, including the entire top-band sloper system and the four HF beams. The principal elevation and azimuthal-plane radiation patterns for the 160-meter array are shown in Figure 6. *NEC* indicates a maximum gain of 3.74 dBi at a take-off angle of 20°, and a front-to-

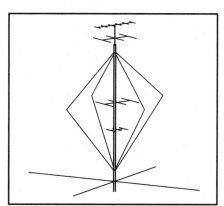

Figure 5—*NEC* model of the 160-meter array as mounted on K3LR's 190-foot tower.

back ratio of 19.10 dB. The driving-point impedance predicted by the computer (for any one of the four elements) is 80.4 + j 65.2 Ω at a frequency of 1.84 MHz. Table 1 shows the *NEC*-generated values for the forward gain and front-to-back ratio when the top-band antenna is mounted on the 190-foot tower, both with or without the HF beams, and with or without the four elevated horizontal radials.

Construction

The four bent dipoles were made from standard #14 insulated electrical wire. The insulation necessitated an increase in the final element lengths in order to achieve resonance at the desired frequency. All of the elements were fed with home-brew

Table 1

NEC-Predicted Performance of the 160-Meter Sloper System in Various Configurations

Description	Forward Gain (dBi)	Front-to-Back Ratio (dB)
No HF beams, no radials	2.88	18.59
No HF beams, 4 radials	3.85	15.59
4 HF beams, no radials	3.27	14.42
4 HF beams, 4 radials	3.74	19.10

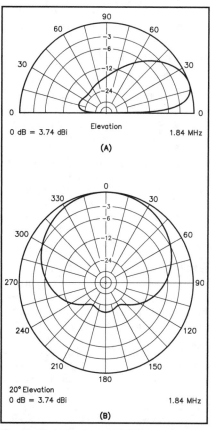

Figure 6—At A, elevation-plane radiation pattern for the K3LR 160-meter array. At B, azimuthal-plane radiation pattern. Maximum gain is 3.74 dBi. *NEC* models ground losses more accurately than *MININEC*-based programs when wires are located close to ground. The elevated radials are 13 feet high. The four antennas have their maximum radiation in the NE, SE, SW and NW directions.

ferrite-bead current baluns, with the length of the balun taken into account when determining the electrical length for each of the Belden RG-8X feed lines. These slopers are oriented at 45°, 135°, 225°, and 315°.

We used a fiberglass box to house the antenna-switching circuitry, so that the coax connectors would be insulated from each other without having to use special hardware. Four double-pole double-throw Deltrol relays (from Surplus Sales of Nebraska)[5] were used for driven-element selection. Five UHF chassis-type female coax connectors were installed on the outside of the fiberglass enclosure, with the 12-V relays mounted inside the box immediately adjacent to the SO-239 fittings. A conventional antenna switch will not work for this project, because both the center conductor and the shield must float, to allow the coaxial feeders to act as loads.

> During the 1993 CQWW CW Contest, the top-band sloper system really shined. With nearly 200 QSOs in 72 countries, the array kept Tim's station close to the top on 160 meters.

Four 15-foot pressure-treated 4×4-inch wooden poles were installed at azimuth angles of 0°, 90°, 180° and 270° to support the 133-foot elevated radials attached to the tower at a height of 13 feet. These horizontal radials are also made of #14 insulated wire.

Testing

When initially constructed in accordance with the original design dimensions, the radiators were a little bit short, since minimum SWR occurred at about 1.86 MHz. Each element was then lengthened by 1.3 feet (for a total of 131 feet, 8 inches) to bring the resonant frequency to 1.84 MHz. At that point, the SWR in the directional mode was so low that no impedance-matching was required.

The array seemed to perform somewhat erratically at first, but it became much more stable after the "common point" was grounded. This was done by running a short, heavy wire from the chassis-mounted SO-239 coax connector for the main feed line (at the Hoffman enclosure) to the base of the tower, which is set in cement and grounded via three 8-foot rods. In this manner, the outer

Table 2
Approximate Dimensions for the Sloper System on Various Amateur Bands

Freq (MHz)	Tower Attach (Feet)	Element Length (Feet)	Anchor Distance (Feet)	Feed Line Length* (Feet)	Radial Length (Feet)	Radial Height (Feet)
1.8	200	131	208	176.2	133	13
3.7	100	65.5	104	87.8	66.5	10
7.1	60	34	62	44.9	34.5	6

* These lengths assume that Belden RG-8X is used, and that a 14-inch length of RG-142 is added at the feedpoint for construction of a ferrite-bead current balun.

shield of the main coaxial feeder is tied directly to a good earth ground immediately adjacent to the relay switch-box.

The antenna system was tested both with and without the four elevated radials, and the front-to-back ratio appeared to be at least 20 dB in either case. The difference in front-to-back ratio, which had been predicted by *NEC* was not noticed, either because it was too small to be discerned, or because other environmental factors in the real world were involved. The array seemed to play slightly better with the radials (perhaps because of some extra forward gain), so they were left in place.

Since the wire elements and the support ropes are very long, there is a fair amount of sag in the system. As a result, the lower ends of the slopers actually overlap at the base of the tower, and the wires extend rather close to the ground. Thus, all four ends are spaced well apart, in order to avoid arcing, which can occur if the wire elements should accidentally touch each other (or the tower itself). This problem could be avoided completely if the anchor-points for the support ropes could be raised off the ground, or moved farther away from the tower base. In addition, the array would fit somewhat better if the tower itself were slightly taller.

Operation

The sloper system was used for the first time during the CQ World-Wide SSB DX Contest in October 1992. K3LR was operated in multi-multi class, and the new 160-meter antenna worked very well. The operators (Alan, N3BJ, and Scott, WR3G) felt that the array was "loud" and that there were no problems hearing or working any station. The transmitting setup was a bit compromised by the use of an old amplifier that put out only 800 W. Even at that power level they completed 124 QSOs, working 13 zones and 32 countries. In the 1993 SSB Contest, the score was 102 QSOs, 12 zones, and 36 countries. All of these numbers stack up very well against other top multi-multi entries.

The sloper system was again put on the air during the ARRL 160-Meter Contest in early December 1992. This time K3LR was entered in the multi-single category with WR3G, W3YQ and K3LR as operators. The final total was 1333 QSOs and 99 multipliers, a new all-time record for this category. During the contest, the big antenna was

compared to an inverted **V** (apex at 150 feet) located 750 feet away from the 190-foot tower. The parasitic array was always one S-unit better than the inverted **V** in a desired direction, as long as the station was at least 500 miles away. However, there were times when close-in stations were better on the inverted **V**.

During the 1993 CQWW CW Contest, the top-band sloper system really shined. With nearly 200 QSOs in 72 countries, the array kept Tim's station close to the top on 160 meters. K3LR worked 104 Europeans, compared with the 111 European stations worked by K1AR. Since K3LR is in Western Pennsylvania (only half a mile from Ohio), the new antenna is the "secret weapon" that enables Tim to be competitive with the East Coast stations.

Interactions

K3LR has noticed some *minor* fluctuations in the SWR readings taken while rotating the lower 40-meter beam (which is mounted at 100 ft), and he attributes these variations to the presence of the 160-meter sloper system. Otherwise, there have been no discernible effects on the remaining HF beams due to the array.

Using the Array on a Different Band

The design can be scaled easily to other frequencies, and suggested *initial* dimensions are given in Table 2. We are looking forward to receiving comments from others who build or modify this antenna system.

Acknowledgments

We'd like to thank Scott Jones, WR3G, and Tim Jellison, W3YQ, for their assistance in building the antenna, and for their operational observations.

Notes
[1] D. Pietraszewski, K1WA, "7-MHz Sloper System," *The ARRL Antenna Book*, 1991, pp 4-12 to 4-14.
[2] D. C. Mitchell, K8UR, "The K8UR Low-Band Vertical Array," *CQ*, Dec 1989, pp 42 to 46.
[3] *ELNEC* is available from Roy Lewallen, W7EL, P. O. Box 6658, Beaverton, OR 97007.
[4] G. J. Burke and A. J. Poggio, *Numerical Electromagnetics Code (NEC)—Method of Moments*, Naval Ocean Systems Center, San Diego, CA, Jan 1981.
[5] Surplus Sales of Nebraska, 1502 Jones St, Omaha, NE 68102. Tel 402-346-4750.

Curtains for You

Big decibels for small dollars.

Last February I placed fifth in the US in my entry category in the ARRL International DX Contest with an antenna system that cost less than a hundred dollars and took just two weekends to erect—a classic Sterba curtain.

Several factors brought about my decision to hang two Sterba curtains in the trees:
- Mother Nature, who gave us the warmest December on record;
- A compulsion to store a couple hundred feet of balanced feed line in the air instead of under the car, where inevitably I would run over it;
- The lack of initiative to erect a tower and beam;
- The ARRL CW DX Contest;
- An article by John Schultz, W4FA, in *CQ* Magazine[1] that said, "This classic antenna . . . develops an awful lot of gain for a relatively compact wire antenna."

For seven years I'd been staring up at the 60- to 70-foot trees surrounding my house and thinking about wires, but my results over the years with directional wire arrays had been dismal. Longer ago than I care to admit I had a 500-foot end-fed wire that laid down a good signal on several bands.

Trouble was, it had a beamwidth of about three degrees.

> ## The Bell System was using Sterba curtains for its long-distance short-wave telephone circuits on the HF bands at least as early as 1930.

Later, I had triband beams and spent a lot of time trying parasitic arrays such as fixed wire Yagis and quads for 40 meters. They refused to work well, though. Usually, when I had a couple of trees to work with, my wire antennas ended up with gain to someplace where penguins outnumber people and hams make expeditions to about once a generation.

I started thinking about Sterba curtains some time before Schultz's article appeared in *CQ*, but he still gets much of the credit for the push I needed. The trees were bare, 10 meters was wide open, and I could get the top of a Sterba up about 45 feet.

A Classic Antenna

If you're interested in a technical explanation of how the Sterba curtain works, see the sidebar "Technical Stuff." It's really a pretty simple antenna.

The ARRL Antenna Book says, in a typical understatement, that the use of Sterba "arrays" by amateurs has been "rather limited."[2] (I prefer "curtain" to "array" because that's precisely what it looks like.) Nevertheless, when I mentioned my project to the boss, Dave Sumner, K1ZZ, he recalled experimenting with a Sterba when he lived in Iowa one summer 20 years ago. (I expect the idea popped into Dave's head after a few days of navigating to work through the Collins Radio antenna farm.)

The Antenna Book notes that the Sterba is a closed loop system for direct current and low-frequency ac, and suggests that "heating currents can be sent through the

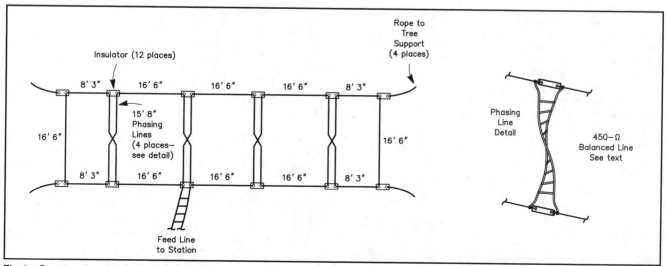

Fig 1—Construction details for an eight-element 10-meter Sterba curtain. Design frequency is 28.4 MHz. Note that phasing lines are twisted once so that the conductors cross. The inner end of an upper element feeds the outer end of a lower one.

Technical Stuff

Fig 1 in the accompanying article shows the layout and dimensions for constructing a 10-meter Sterba curtain or Sterba array (either name is appropriate). The construction is simple enough, but how do all those wires work together to make such a hot antenna? Actually, it's pretty simple. Just consider all the horizontal wires as a series of half-wave dipole sections, and all the vertical wires as phasing lines.

To understand this, let's first consider two parallel half-wave dipoles stacked one above the other, with a separation of ½ wavelength. Let's also feed the two dipoles in phase, for maximum performance in the broadside direction. Fig A shows such an antenna arrangement (ignore the broken lines at the ends of the dipoles for a moment). The feeder brings power to the bottom dipole, and a phasing line carries power from the feeder to the top dipole. Because there is a half-wave between the two dipoles, and thus a ½-wavelength phasing line, it takes half an RF cycle for the phasing line to carry power from the system feed point to the top dipole. If the phasing line were simply connected there without twisting, the two dipoles would then be fed 180° out of phase. But we

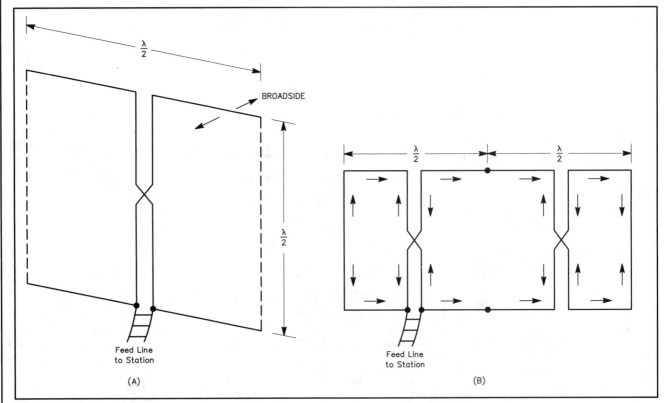

At A—Two half-wave dipoles fed in phase (ignoring the broken lines). Adding wires at the ends (broken lines), does not alter the phase and helps to equalize the power radiated from the two elements. This is the basic Sterba antenna section. At B—A simple 4-element Sterba array containing two sections as shown in Fig A. The arrows show the direction of current flow throughout the array at a given instant. Radiation from the vertical wires tends to cancel, while that from the horizontal wires reinforces.

wires to melt the ice that forms in cold climates." This piece of information led me to suggest to a friend the possibility of plugging the feed line into a house outlet (keeping a fire extinguisher handy by the circuit breaker, of course). But I digress.

It seems that in typical ham style, the Sterba array was in use before it got the name or even before appearing in a scientific paper. The Bell System was using Sterba curtains for its long-distance short-wave telephone circuits on the HF bands at least as early as 1930. E. J. Sterba, who worked at the Bell Telephone Laboratories in New York City, described the array in the July 1931 issue of the *Proceedings of the Institute of Radio Engineers* (starting on page 1184).

Sterba, in an article titled *Theoretical and Practical Aspects of Directional Transmitting Systems*, omitted "details of the mathemati-

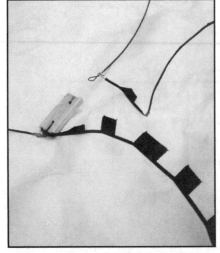

Fig 2—So much for insulators made from furring strips.

cal derivations . . . for brevity." Regardless of the brevity, I still couldn't understand the math (which is jam packed with weird Greek letters) but there, on page 1203, is a diagram of my antennas.

Sterba diagrammed several wire arrays, including what are now called the "Lazy H" and the "Bruce Array" (also described in Chapter 8 of *The ARRL Antenna Book*). The array we now call the Sterba curtain was the antenna in use at that time at the Bell System's Lawrenceville, New Jersey, facility.

I believe that Bell settled on the Sterba configuration for exactly the same reasons that I find it so compelling: it's easy and cheap to build and to keep up, and it just looks right.

A few years later, John Kraus, W8JK, in his classic book *Antennas*, mentioned curtain arrays, and apparently coined the term

don't want that. Rather, we give the phasing line a half twist, as shown in Fig A. This inverts the 180° phase difference, and the two dipoles are now fed in phase. A simple way to look at it is to consider that the half twist is necessary to compensate for the time delay in the phasing line.

We can now connect single wires between the ends of the dipoles, as indicated by the broken lines in Fig A. In doing that, what you have now is your basic Sterba antenna section.

Extending the Array

It has already been said, "The system obviously can be extended as far as desired." Actually, the antenna of Fig A by itself, with the vertical end wires, will do a very respectable job, but extending the array provides even better performance. This is done by adding more Fig A Sterba sections, simply stringing them together end to end.

Fig B shows a Sterba array containing two sections (four half-wave elements). Note that the added section, to the right in the drawing, is not fed independently. Instead, the single vertical wires are not installed at the junction of the two half-wave sections; the added section receives power through its direct connection to the dipole ends of the original section. This is possible because the currents in all the dipole elements of the array are flowing in phase, as indicated by the arrows in the drawing. Logically, an extended array should be fed as near the center as possible, to preserve symmetry and thereby maintain current balance in the feeder. The system may also be fed by opening and feeding a corner. This is sometimes done at a bottom corner with coaxial line and a toroidal step-up transformer. For either type of feed, the feed-point impedance is in the order of a few hundred ohms.

So now you can see that Fig 1 in the accompanying article is really nothing more than several Fig A sections connected in series—four such sections, to be exact. So the author properly refers to this antenna as having eight elements; each is a half wave in length.

Radiation Pattern and Gain

What kind of a radiation pattern and gain does the Sterba array have? Fig C shows the results of an analysis using a computer program based on MININEC, which performs calculations using a method of moments. The theoretical gain is approximately 8.1 dBd, and the 3-dB beamwidth is 24°. Only the horizontal wires and the two vertical end wires were included in the computer model, so this is not a totally accurate analysis. In the real world, the beamwidth will likely be somewhat greater, the gain somewhat less, and the nulls of the pattern filled in slightly. Radiation from the interior phasing lines will contribute to these effects.

Whatever the case, you can be sure of one thing—a Sterba

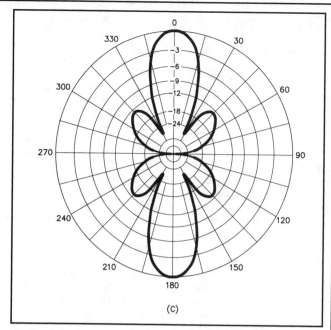

Fig C—Azimuth radiation pattern for the antenna of Fig 1, obtained from computer calculations using a *MININEC*-based program. The plane of the antenna is along the 90-270° line.

curtain works, and it works very well! I can vouch for that based on personal experience. Many years ago I was at a field site doing work on a propagation study. We had a 1200-ft-long rhombic antenna we used on occasion for 22-MHz communications with our net control station, roughly 5000 miles away. It was a time in the sunspot cycle when conditions were terrible, requiring a superb antenna. But the rhombic was dedicated for use in transmitting data most of the time, and that left us in need of a good antenna for point-to-point communications. We found a high dipole to be almost useless. In desperation, we hung a 12-element Sterba curtain between the side supports of the rhombic. Its upper bay was 70 feet 1½ wavelengths) above ground. The Sterba design came right out of *The ARRL Antenna Book*, but the book didn't give much of a clue about how it would perform. However, its performance was truly amazing, almost the same as the rhombic! And that for only a few hundred feet of wire, some rolls of open-wire line, and some insulators. You bet a Sterba curtain works!—*Jerry Hall, K1TD*

Sterba curtain. In 1939, the first edition of *The ARRL Antenna Book* noted the Sterba array. Half a century later the information is basically the same.

My favorite line in *The ARRL Antenna Book* is "The system obviously can be extended as far as desired." This was what sent me into the balmy 50-degree December weather, to pace off distances between trees and plot paths to Europe, Asia, and New Zealand.

Preparations

Here are the highly technical and complex set of instructions for erecting your Sterba curtain:

A. Build the antenna.

B. Put it up.

If this seems disarmingly simple, it is. Actually, there is one thing to do before anything else, and that's to determine

whether you have the room for this project.

The first Sterba I built, eight elements, is shown in Fig 1. It's roughly 64 feet long—about the same as a 40-meter dipole. You can add or subtract elements, depend-

> *I still couldn't understand the math (which is jam-packed with weird Greek letters) but there, on page 1203, is a diagram of my antennas.*

ing on the amount of room you have. As for height, my curtains at 45 feet work great. I suspect they would work with the lower wires as close to the ground as perhaps 10 feet. The literature says the bottom of the array should be a half-wavelength above ground—that's only 16 feet at 28 MHz.

If you are lucky, you will be able to support two curtains at right angles to each other from three trees or other supports.

Materials

Construction materials: antenna wire, balanced feed line, insulators. These are the same materials you need to build a dipole, only you need more for a Sterba array.

I made a marvelous discovery in the evolution of my curtains. For some time I'd wondered where to get antenna wire these days (without resorting to mail order). I would rather take up another hobby than use the kinky, copper-clad steel stuff available today on every other street corner. You know, the "antenna wire" that comes in 50-foot rolls only, that turns black after a week, and which the store is always out of anyhow.

Then I dropped by the local hardware store and was browsing the wire aisle. Voila! "#14 solid black 500 feet." I almost had an accident. Finding this wire was for me akin to being the inventor of the internal combustion engine and walking into a dry goods store to find spark plugs for sale.

The other wire needed for this project is balanced feed line. "Open-wire," we used to call it. Don't make your own.

I like the black, plastic-insulated, 450-ohm variety. This is a product I always have figured the manufacturer would discontinue for lack of demand, and I collect it like the proverbial ball of string. Last winter, I helped take down somebody's tower. When I went back in the spring to clean up ("the ice that forms in colder climates"), I unearthed an end of this very same balanced line. Down on hands and knees archaeologist-style, I delicately pulled the line from the dirt and grass. And pulled, and pulled. It was the holy grail of feed lines, nearly 200 feet, all one piece, dirty but definitely serviceable. I hosed off the whole mess and threw it into the washing machine on "heavy load."

Finally, insulators. Since Sterba curtains need a lot of insulators (12 for my 8-element array), I decided to make my own. I had some 1.5-inch-wide furring strips (¼-inch-thick wood) lying around, so I cut them into 4-inch-long pieces, drilled two holes in each one, and that was that. These were meant to be very temporary, just to see if a Sterba curtain would work. Of course, there's a lot of pull on these insulators, especially along the top of the antenna, and I estimated they would last about a month.

It was after about two months, actually, that they began failing in a spectacular manner. You will note that the insulators are used where the dipoles attach to the balanced feed line. When an insulator fails, the only thing holding the connections together is the plastic strip making up the spacer for the balanced line. As the wind, the "ice that forms in northern climates"— or whatever causes the failure—persists, the balanced line splits down its length (Fig 2). The curtain loses its poise and things become an ugly mess in a hurry.

So do it the right way, or maybe even the way I switched to. I bought some ¾-inch

diameter hardwood dowel, cut it into 4-inch-long pieces, drilled holes in it, and then boiled the pieces in ordinary canning wax. I honestly don't know if waxing does any good, but the books have recommended it for at least a century.

I was doing some boiling one afternoon when a friend stopped by and reminded me that you aren't supposed to put the pan of wax directly on the stove top. So Switch to Safety and use a double-boiler. I didn't think to try microwaving the wax.

Build the Antenna

Now build your Sterba curtain. You can assemble a curtain in a garage or basement but I don't recommend it. Ever try to relocate a spider web?

Rus Healy, NJ2L, covered a lot of construction tips for wire antennas in his magnum opus on dipoles.[3] You might want to review those articles before starting on your curtain. [These articles can be found elsewhere in this book.—*Ed.*]

Dimensions for an eight-element, 10-meter Sterba are shown in Fig 1. Measure one dipole, then cut all the others to the same length. Remember to leave enough extra wire to wrap around the insulators. Do the same for the phasing lines. Although the phasing lines are also ½ electrical wavelength, they're physically shorter than the dipoles to account for the velocity factor of the 450-ohm balanced feed line (about 0.95 for the line I used).

It's a good idea to consider soldering the connections, too. Since you probably will be building this in the dead of winter, think about measuring all the dipoles and phasing lines inside, tinning all the wires, then moving outside for assembly. On really cold days even giant electric soldering irons strain—those nifty little butane-powered irons that you can buy at Radio Shack and elsewhere are a good alternative.

These antennas are not critical in terms of dimensions. Although I did measure the dipole elements carefully in the beginning, each time I replaced an "insulator" I lost about six inches of dipole length. I also somehow ended up with a couple of phasing lines perhaps as much as two inches too long. I'm sure there is a limit to how much of this you can get away with, but Sterba curtains seem to be very forgiving.

Put It Up

Let's assume you are using trees for supports. Much has been written on this subject. Methods suggested for getting ropes into trees include, of course, climbing them. I rule out this method because you want to get the rope as high in the tree as possible, which means up in the tiny branches where you can't climb.

Mechanical methods described over the decades run from bow-and-arrow to slingshot[4] to catapult. I know for a fact that none of these methods work because I have never tried them.

I went to a fishing supply store and bought some 6-pound-test line and a hand-

ful of sinkers. Then I bought a ball of small nylon twine. The big expense was for hundreds of feet of polypropylene rope (that, despite rumors, will last quite a while holding wire antennas).

I used the horseshoe-pitching method, swinging about three feet of the fishing line until I got the right momentum, then letting go. There's an art to this, and if you can't master it, use a slingshot or some other method.

The fishing line disappears once you reel it out onto the ground. It can easily become a hopeless bird's nest of knots and will break at the slightest provocation. I lost several sinkers and a lot of the fishing line on unsuccessful launch attempts, but the beauty of the stuff is that it's invisible up there in the trees and it's cheap.

The spectacle of a grown man or woman flinging fishing sinkers into trees can invite a visit by the attendants in white coats, so it might be prudent to do this when the neighbors are gone.

Now use the fishing line (you already know this) to pull up the nylon twine, then pull up your sturdy rope. You can reuse the twine to tie off the lower corners of your Sterba (see below). In the spring your lawn mower will easily locate the lost fishing line.

Keep in mind that in time the motion of the branches may saw through your big rope (or the tree may be hurt). Every few months you can reposition your curtain (and the ropes). I, of course, accomplished this by lowering my curtains several times to replace those furring strips.

Just as a clean car runs better, so a straight and level curtain works better. When you pull your curtain into position you will be overjoyed to see everything fall neatly into place. These are very elegant, graceful arrays. And the real beauty is that you can pull on all four corners. You can play with ropes and tension until all is shipshape.

Even if your top supports are at different heights, your curtain will find its own spot and level off nicely. I found the most rewarding part of putting up my curtains was being able to fine-tune them for esthetics. I absolutely will not climb a tower to reposition one Yagi element that is a degree off level, but readjusting ropes is easy and effective.

Feeding the Array

You'll need an antenna tuner that can handle balanced line. The Sterba is a beautifully symmetrical, electrically balanced antenna that tunes like a dream, and one setting of your antenna tuner covers at least a hundred kilohertz.

You can feed a Sterba with coaxial cable if you use an appropriate impedance transformer, but I don't recommend it. For the price of the coax you probably can buy or build an antenna tuner, and with coax you lose the ability to use the Sterba on other bands (more on this later).

I was fortunate to have two antenna tuners available, one for each Sterba. One tuner could be used with two Sterbas; you would need a way of switching the feeders to the two antennas (are double-pole, double-throw knife switches still made?). You also would want to make the Sterbas electrically identical so the tuner wouldn't have to be readjusted.

First Impressions

I started out with an eight-element Sterba aimed broadside at about 70/250 degrees (remember, Sterbas are bidirectional). It cooked from the moment I put power to it, the first contact being with ZD8LII on December 21 at 1250Z (a "band opener"). Three days later I tried it on 24 MHz—tuned up just like it did on 10 meters and blew a hole in the 12-meter band.

In a couple of weeks I was starting to think seriously about a single-band 10-meter effort in the ARRL CW DX Contest in February. Around Christmas, a ham friend from Austin came to visit. After plying him with a couple of Wild Turkeys, I dragged him out into the balmy 40-degree weather to observe another fishing line/sinker session, into the two trees I had scoped out for a north-south Sterba.

I built the second Sterba one evening that week, finishing up after dark, soldering by

A tribander on a 60-foot tower will cost a lot more and definitely is not neighbor-friendly.

kind of aerial. I moved several of these stations down to 12 meters and worked them there, too. But the best was yet to come.

With the contest just two weeks away, I lowered the east/west Sterba curtain and added four more dipole elements, making it about 90 feet long. I managed to reorient it to about 60/240 degrees, still some 20 degrees off optimum for Europe, but better than before. When you begin making antennas bigger at this level, the differences are difficult to discern. But the antenna was a lot bigger, giving me a psychological boost.

On the other hand, the 10-degree reorientation seemed to pay off. I did better into Scandinavia, and 4K2/UV3CC on Franz Josef Land gave me an S9 + 30 dB report.

These antennas seem to have very broad patterns (in two directions, of course). That's good, because neither of my Sterbas is optimally oriented to hit the high-population areas of Europe and Japan. The north/south antenna, at 5/185 degrees, from my Connecticut location, is about 20 degrees off for Japan and most of South America. You can't move the trees....

I believe these antennas have all the advantages noted for cubical quads—namely, they are quiet, and are great listening antennas because of their large capture areas.

Sterba has just over 200
counting the phas-
ave the disadvan-
s off the back. But
where this is not
meters.

Sterbas has been
ice on 12 meters;
"textbook" there
small) but I break
many people have
ters yet.

:pt power on every
to 28 MHz. They
t of gain on 20, 17
hard to tell where.
et out about as well
t center-fed flattop.

The Ultimate Test

In the days before the contest, I began listening to the band every morning, noting the appearance of the first Europeans. Ditto in the evening for Japanese stations on the other curtain.

I operated the contest using my curtains and a 600-watt amplifier. In the end, my score was good for fifth place in the singleband 10-meter (CW) category. Stations who beat me used stacked Yagi arrays. Although I had higher contact totals than several of the multi-transmitter stations, I lost the contest on multipliers (number of different DXCC countries). It is obvious that my curtain antennas have deep nulls off their ends. I worked only one African station, J52US, over the north pole late at night. I also worked almost nothing in the Pacific, another null in my system.

What I needed was a third Sterba or a small rotary Yagi or quad to "fill in the gaps." In early June I found the initiative to put up 60 feet of tower and a three-element triband beam. Thanks to the usual lousy summertime conditions there has been little opportunity to compare antennas. On a couple of sporadic-E openings to Europe, the tribander and Sterba were neck-and-neck. Into Central and South America, sometimes one or the other is significantly better. On summertime sporadic-E openings down the East Coast, the tribander almost always is a little better than the north/south Sterba for stations in the Carolinas, Georgia and Florida.

If the Sterbas are about as good as a beam, why not go with the beam? My two Sterbas cost less than a hundred dollars and are virtually invisible. A tribander on a 60-foot tower will cost a lot more and definitely is not neighbor-friendly. On the other hand, the Sterbas have reminded me of what amazing performers small tribanders are, considering their design limitations.

Here's an experiment for someone to try— a Sterba made for midway between 15 and 10 meters; that is, for 12.5 meters. I know it would work beautifully on 24 MHz, and suspect it also would rival a tribander on 10 and 15.

These arrays are so simple to build and erect, perhaps I'll try. Then again, why don't you?

Notes

[1] J. Schultz, "A New Look at Some Classic Wire Antennas," *CQ*, Jan 1991, pp 32-34.
[2] J. Hall Ed., *The ARRL Antenna Book*, 16th ed, pp 8-41 and 8-42.
[3] R. Healy, "Antenna Here is a Dipole," *QST*, Jun 1991, pp 23-26 and "Feeding Dipole Antennas," *QST*, Jul 1991, pp 22-24.
[4] W. Calvert, "The EZY Launcher," *QST*, Jun 1991, pp 34-35.

Bidirectional Antennas for Field Day

Are you working everyone you can from your Field Day station? Here are some antennas that may improve your odds!

Everyone who has operated Field Day has experienced times when it feels like you've "worked 'em all," and that there are no new stations to contact. You call CQ repeatedly with no replies. Is the problem related to a lack of activity? Or have you really equipped your station with the right antenna(s)?

This article provides planning and construction information for a time-honored class of antennas which can "spray" your FD signal in more than one direction, while still providing competition-grade gain. As an additional benefit, several of the antennas described provide deep side nulls, which can be useful in minimizing inter-station interference in multi-transmitter Field Day operations.

Population Density and Antenna Patterns

It is well known that the eastern states and Canadian provinces contain the most widespread area of high population density in North America. Yet California has more than twice as many amateur licensees than any other state (over 100,000!). Significant

> *What you need for maximum scoring success in Field Day is the widest exposure possible to the most amateur population with the loudest possible signal.*

amateur concentrations are also found in northern Oregon, Washington, and British Columbia.

But if you're in the Midwestern part of the continent, what sort of antenna did you use, say, on 10 meters last year? A three-element Yagi? And which direction did you point it during Sunday morning's sporadic-E opening? Probably toward South Carolina, with its 8500 friendly licensees, at the expense of all the potential QSOs toward the west. What you need for maximum scoring success in any contest or Field Day effort is the widest exposure possible to the most amateur population with the loudest possible signal. And unless you're operating from a space shuttle or *Mir*, this probably means that you'll need to spray your RF energy in more than one direction at a time.

The problem isn't just one of signal strength in the rear direction. Activity patterns on the bands are also important. One might conclude that, if you're S9 in the front direction with a beam that has 24 dB of front-to-back ratio, your signal off the rear will be S5, still enough to make a contact. On Field Day, though, clubs and groups operating in classes like 1A (single transmitter) have to make hard decisions about the band on which they operate at any given time. If the operators of a California club only hear S5 signals from the Midwest on 10 meters, but S9 signals on 20, chances are that they'll stay on 20. If the poor W6 *does* go to 10, it doesn't take too many times calling a station in Arkansas with no response (because the W5 is working S9+ *East* Coast stations) to cause the 6-land operator to QSY back to 20!

Opportunities for the utilization of bidirectional antennas are not limited to the highest bands, either. On 20 meters, for example, a station in the middle part of the East Coast (say, North Carolina) might want a three-element Yagi for working to the west, but a bidirectional antenna for simultaneously working stations in New England and Florida on those big Sunday morning E-layer openings which

always seem to occur on Field Day. A dual antenna system such as this, with quick-switch capability, will be a formidable weapon in the competitive aspect of Field Day, while demonstrating the benefits of multiple antennas for emergency communication planning.

Antenna Design Examples for Increased Coverage

One of *my* favorite Field Day antennas is the "Bisquare." An easy-to-build variant of the "Lazy-H" antenna, it can be built out of wire and, for frequencies above 21 MHz, be erected on a lightweight mast of about 40 feet in height. The Bisquare is easily tuned by a wide-range balanced-feed antenna tuner; for both efficiency and sentimental reasons, I use the old Johnson "Matchbox" tuner for this purpose.

Despite its appearance (see Figure 1), the Bisquare is not a "loop" antenna, as it is *open* at the top. It produces a horizontally polarized signal, which at moderately low heights, emits broad signal lobes that can illuminate a wide

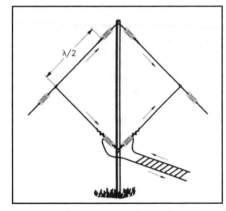

Figure 1—Basic layout of a Bisquare antenna. Note that the top corner of the antenna is open-circuited.

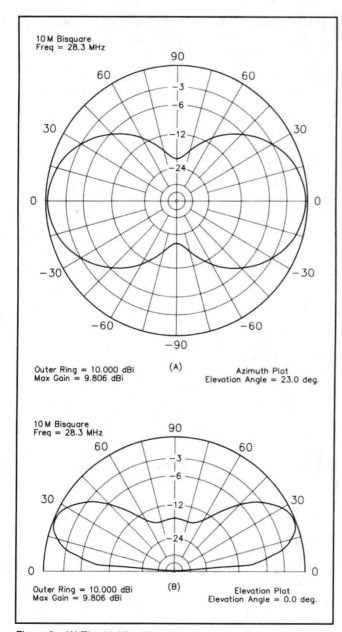

Figure 2—(A) The 28-MHz Bisquare azimuth pattern; apex at 30 feet. (B) Bisquare elevation pattern; apex at 30 feet.

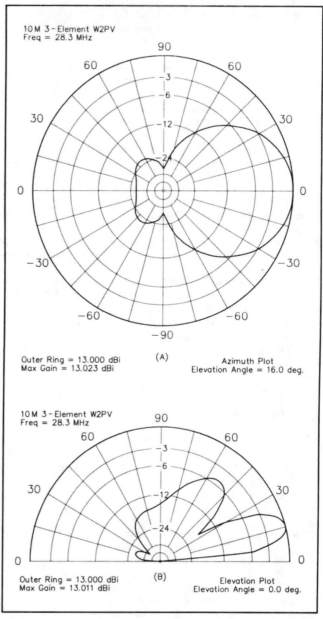

Figure 3—(A) Three-element Yagi 28-MHz azimuth pattern; boom at 30 feet. (B) Three-element Yagi elevation pattern; note the high-angle lobe.

geographical area. Figure 2A shows the azimuth (horizontal view) pattern for a 28-MHz Bisquare with its apex at a height of 30 feet, while Figure 2B shows the elevation (side view) of main lobe) pattern. For comparison, Figures 3A and 3B show the corresponding patterns for that most typical of all Field Day antennas, the three-element Yagi. In analyzing these patterns, we need to look at two important factors:

1) How does our signal compare in strength to our reference antenna (the above-mentioned three-element Yagi) in its favored direction?

2) What additional coverage area, if any, do we pick up with the (bidirectional) Bisquare compared to the (unidirectional) Yagi?

Comparison of Figures 2 and 3 shows that the three-element Yagi has a peak gain (over "real" ground) of 13.0 dBi, while the Bisquare is somewhat lower at 9.8 dBi (the *apex* of the Bisquare and the *boom* of the Yagi are both assumed to be at a height of 30 feet, or

about 0.87 λ). The front/back ratio (F/B) of the Yagi, however, being 27.6 dB, means that the Bisquare has a gain *advantage* of 24.4 dB in the rearward direction of the Yagi. If we assume that the Yagi is producing an S9 signal somewhere in its peak forward direction, this means that the Bisquare's signal will "only" be S8.5 (assuming 6 dB per S unit). Moreover, the 27.6 dB F/B of the Yagi means that the Yagi's signal at 180° away from its peak will be about S4.5, *while the Bisquare's will still be S8.5!* Small rear lobes of the Yagi not in the 180° position mean that, in some directions, the difference will not be quite so dramatic; on the other hand, even deeper nulls in the Yagi's pattern mean that the difference may be *greater* on some headings. Note also the rather large high-angle lobe from the Yagi in Figure 3B. Since this angle often cannot be refracted via the ionosphere on 28 MHz, this signal power may disappear into outer space as wasted energy.

What does all this data mean for a Field

Day group in Oklahoma City? With a three-element Yagi on 28 MHz at a height of 30 feet, they might illuminate an E-layer path which includes the population centers in the W3, W4,

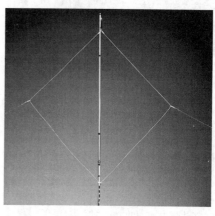

A Bisquare antenna.

and W9 call areas without falling outside the –3 dB (½ S-unit) boundary in the azimuth pattern. But with the Bisquare, the –3 dB azimuth points now include W1, W2, W3, W4, W8, and W9 in the easterly direction, *plus* W6 and southern W7 to the west. The added westerly contribution of 150,000+ amateurs (in Arizona, California, Nevada, Oregon and Utah) means the opportunity for many more QSOs in the log.

To provide coverage of the Chicago, Minneapolis/St. Paul, Toronto, and other due-north metropolitan areas, a northward-pointing Yagi could be mounted on its own mast at about 26 to 30 feet, or two Bisquares may be suspended from a single mast, and fed independently; there will be minimal interaction between the two arrays. A simple coaxial switch will allow quick direction change.

Building a 10-Meter Bisquare

Unlike most parasitic antenna arrays (such as the Yagi or quad), the dimensions of the Bisquare are entirely noncritical. The Bisquare consists of two one-wavelength wires, formed into a diamond, and fed at the bottom. From the formula L=984/F (MHz), we find that a length of 34 feet 9 inches produces the proper length at 28.316 MHz—perfect for Field Day operation! Cut two such wires, leaving an additional 4 inches on each end for looping through the top and bottom insulators.
through the top and bottom insulators. Find the center point of each wire, and install an insulator at that point. Now install insulators at the "top" end of each wire (see Figure 4 for details on a simple method of insulator installation), and at the bottom end. A small Dacron-cord loop may be tied to the top insulator pair, and Dacron cords may be tied to the center insulators so that they may ultimately be pulled outward to form the final (square) shape.

Install a pulley on top of your mast, and tilt the mast to the vertical position. Once the guy ropes (nonconductive material is essential!) are secure, pull up the two wires to the maximum height. Now it's time to attach the feed line. The feed line is of the open-wire type. Commonly available "ladder line" will also work just fine in this application. The open-wire feed line is soldered to the bottom ends of the two wires, and is allowed to hang downward so as to reach the antenna tuner. In this application, we do not worry about the high SWR that may be present on the feed line, as the extremely low loss of the open-wire line means that the *additional* loss due to SWR will be negligible (for an extensive discussion of this subject, see Chapter 24 of *The ARRL Antenna Book*, 1997 edition). Once the feed line is attached, you may pull outward on the lines attached to the center insulators.

When tuning up an antenna of this type at the Western Amateur Radio Association's N6ME Field Day site, I use an "Antenna Analyzer" manufactured by MFJ Enterprises which allows me to adjust the Johnson Matchbox for a perfect match in less than 30 seconds *without* the need for a generator to power a transceiver (similar products are also manufactured by AEA-CIA and Autek). When adjustments have been completed, the antenna tuner is wrapped inside a plastic garbage sack to protect it from the weather (it *always* rains on Field Day!), and I'm on the air!

The Bisquare is not the only bidirectional antenna type that may be useful for Field Day applications. Let us now explore some other antenna types, each of which has its own advantages and disadvantages.

The Kraus (W8JK) Flat-Top Array

This simple bidirectional array was invented by Dr. John Kraus, W8JK, and is described extensively in his book *Antennas*; it also is documented in most all amateur texts, including *The ARRL Antenna Book* (Chapter 8 of the 18th edition). The Kraus Array consists of two closely spaced dipoles fed out-of-phase via a crossed phasing line. Open-wire line is then used to feed the array, as with the Bisquare. Figure 5A shows the basic layout of a typical Kraus Array.

This versatile antenna may be built with half-wavelength elements (known as a Single-Section 8JK), one-wavelength elements (Two-Section 8JK), or even longer elements. However, going beyond the two-section model will narrow the beamwidth excessively, confounding our mission to "spray" our signal over a greater area. Depending on your location, you probably will need to use the single-section Kraus Array to achieve the necessary beamwidth; even so, this antenna at a height of 30 feet produces a peak 28 MHz gain of 10.8 dBi, which is about 1 dB greater than that produced by the Bisquare (with its apex at the same 30-foot height as the 8JK).

A major advantage of the Kraus Array is that it is a horizontal-plane antenna; it therefore requires less mast height to achieve the same takeoff angle as the Bisquare. Another advantage is the excellent null at 90° off of each main lobe (Figure 6). If this null is directed at the station (on the opposite mode) on the same band at your Field Day site, overload problems may

Figure 4—Typical rigging of the top corner of the Bisquare antenna. The top of the Bisquare must be open-circuited.

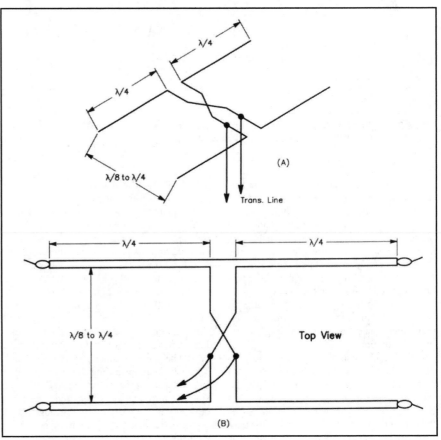

Figure 5—(A) The single-section Kraus Array. (B) A single-section Kraus Array constructed with folded dipole elements will provide greater bandwidth without retuning.

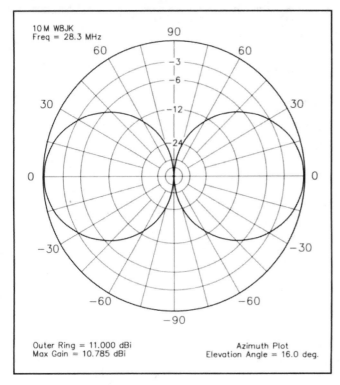

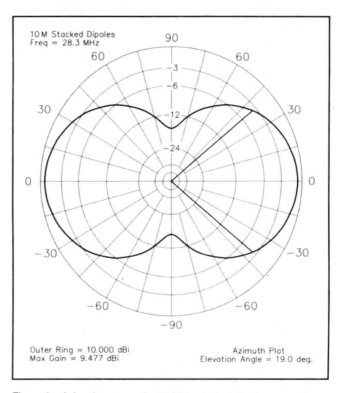

Figure 6—The azimuth pattern of a 28-MHz single-section Kraus Array at a height of 30 feet.

Figure 8—Azimuth pattern of a 28-MHz stacked-dipole array with the top element at a height of 30 feet. Note the broad azimuth coverage area.

be significantly reduced. A disadvantage of the Kraus Array is that it requires two supports, unlike the Bisquare; a Kraus Array may easily be suspended between two masts or trees, however, and the dimensions on 14 MHz, 21 MHz, or 28 MHz are quite reasonable (even a 40-meter version is not out of the question; two standard-length dipoles spaced about 18 feet will do the trick).

The Kraus Array tends to be a fairly high-Q antenna, which means you may need to restrict your operating frequencies (unless you can reach the controls of your antenna tuner). The single-section version may benefit from the use of "Folded Dipole" elements, which raise the impedance and broaden the response. A typical example of this type of construction is shown in Figure 5B.

Stacked Dipoles

Two half-wavelength dipoles may be stacked and phased, as shown in Figure 7, to produce a very effective bidirectional beam. A vertical spacing of one-half wavelength allows the array to be fed in the center (with no transposition of the phasing line, as shown in Figure 7), or at the bottom, with the two dipoles being phased by a crossed piece of open-wire line. At half-wavelength spacing, this array produces a bidirectional gain of 9.5 dBi. With the top element at a height of 30 feet, the bottom element of a 28 MHz array will be at about 12 feet, and this antenna's main lobe will only be down about three-tenths of a dB compared to the Bisquare.

A particular advantage of this type of array is the very broad horizontal lobe, since the vertical stacking produces the gain of this antenna. This means that you can illuminate a huge geographical area within the (80-degree-wide) –3 dB points (see Figure 8). Another advantage stems from the very low RF voltages present at the centers of the dipoles; simple split-dipole elements may therefore be used.

If you have some aluminum tubing available, and can construct a suitable center mount for a dipole element, this is a terrific antenna that is easy to erect. It may also be made using wire, utilizing ceramic end/center insulators, if you have two masts available. Although the gain increases slightly if the spacing is increased to $^5/_8$ λ, the gain advantage is less than 1 dB, making pointless any additional effort at providing this wider spacing.

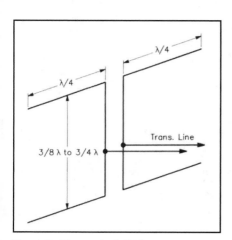

Figure 7—The stacked-dipole array.

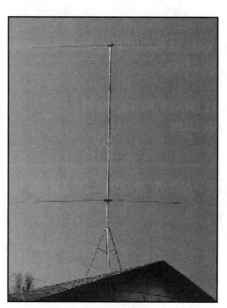

A stacked-dipole array perched atop a roof tripod.

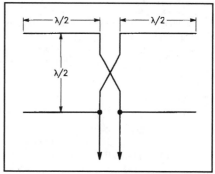

Figure 9—The Lazy-H antenna. Note that the elements are twice as long as the elements of the stacked-dipole array in Figure 7.

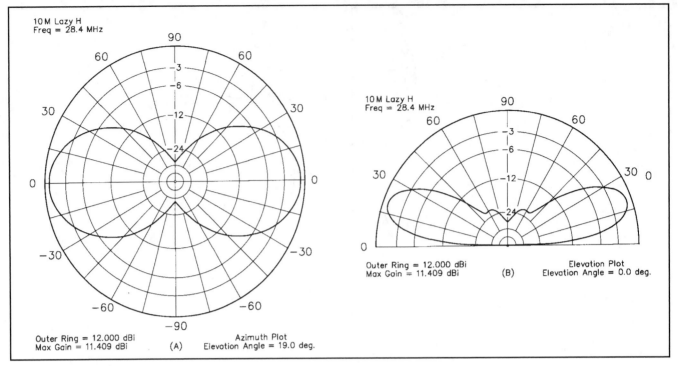

Figure 10—(A) The azimuth pattern of a 28 MHz Lazy-H Array with the top element at a height of 30 feet. (B) The elevation pattern of the Lazy-H.

The Lazy-H

Similar in appearance to the "stacked dipole" array, the Lazy-H is an array of *four* half-wavelength elements, stacked two-over-two to achieve gain via collinear *and* vertical stacking (see Figure 9). Although the gain of the Lazy-H is about 2 dB greater than that of the stacked dipole array, the Lazy-H is of reasonable dimensions for Field Day erection on the 28 MHz band (total "wingspread" of about 35 feet) if suspended between two masts. If tall trees are available, a Lazy-H for 14 MHz or 21 MHz can easily be erected. When half-wave spacing is used, the Lazy-H may be fed at the bottom, with a transposed feed line connecting the top and bottom sections to establish the required in-phase currents.

The gain of the Lazy-H is approximately 1.6 dB greater than that of the Bisquare, and the horizontal pattern is about 10° narrower. Because these two antennas fundamentally are quite similar (the Bisquare being a Folded Lazy-H), the choice of which antenna to use may boil down to a constructional issue. Figure 10A shows the azimuth pattern of the Lazy-H for 28 MHz (top at 30 feet), while Figure 10B shows the elevation pattern. Note the very clean elevation pattern, with no power wasted in high-angle lobes. This is really a great antenna, if you have the supports to get it up.

How Do They Work?

I have used a number of these antennas with good success over the years. On 15 meters from the N6ME FD site, for example, W7 and VE7 stations that were inaudible on a four-element Yagi aimed to the east were a solid S6 on a simple Bisquare positioned to favor the north-south paths. On 10 meters, a similar antenna has occasionally provided *stronger* E-layer signals than a (higher) Yagi pointed the same direction, due to a null in the Yagi's elevation pattern at a critical angle.

While testing a "stacked dipole" antenna at my home in the preparation of this article, I was listening one afternoon to 10 meters toward the East Coast on my six-element monoband Yagi at 70 feet. I heard traces of a weak signal that I could not identify in the local noise; rotating the Yagi back and forth about 45° did not peak the signal. Switching to the stacked dipole array, which was broadside to the eastern USA and mid-Pacific regions, I discovered that the signal was from V73AT in the Marshall Islands, at a solid S7! Although the V73 peaked at between S8 and S9 when the long, high Yagi was pointed at him, one can easily see how a simple bidirectional antenna can really open the door to many contacts that might otherwise be missed!

Summary

Many centrally located Field Day groups make the mistake of considering only a (unidirectional) Yagi-type beam antenna. To change directions, either an electro-mechanical rotator must be installed (something else which can break down!), or someone must run outside and turn the antenna manually. With summer sporadic-E propagation changing so quickly, it frequently is impossible to catch every opening. And from a location in the midsection of North America, there is a very good chance that you will experience simultaneous propagation to the East and West Coasts, a situation that demands a bidirectional antenna capability.

The antennas described previously are simple to construct, cost very little, and require minimal take-down time (a *very* desirable feature on Sunday afternoon!). Because they are "force-fed" via an antenna tuner, adjustment time is measured in seconds; how many of us have wasted *hours* fiddling with a Yagi's gamma match as the sun goes down Friday night before FD? Two bidirectional antennas, erected at right angles, can provide coverage of most of the continent's population centers, with *instant* direction switching. All this at a sacrifice of *perhaps* one-half of an S unit compared to a Yagi's peak gain, while you pick up a *gain* of *several S units* compared to the Yagi's rearward direction. Moreover, in the true spirit of Field Day, the building of high-performance bidirectional wire arrays teaches one the tricks of the trade in setting up an emergency station quickly.

I hope that you will build and enjoy one or more of these antennas in conjunction with your club's Field Day event. You'll be pleasantly surprised at the results!

Resources

John D. Kraus, (W8JK), *Antennas*, New York, McGraw-Hill Book Co., Second Edition 1988 (especially pp. 454-459).
R. Dean Straw (N6BV), Editor, *The ARRL Antenna Book*, Newington, CT, The American Radio Relay League, Inc., Eighteenth Edition 1997.

CHAPTER SEVEN

VERTICALLY POLARIZED

By Dana Atchley, Jr., W1CF

From *QST*, July 1979

Putting the Quarter-Wave Sloper to Work on 160

Want a 160-meter signal that has real DX capability? This half-sloper antenna will put your station in the heat of competition. The cost is next to nothing!

Many of us older amateurs have notused the 160-meter band for years. The reason lies not in a lack of interest. Rather, our failure to participate in 160-meter activities stems from the fact that the typical ssb transceiver manufactured in the United States from the late 1950s to the middle 1970s did not provide 160-meter coverage.

In the past two years, more than just a few of us have traded in our tried and true transceivers that had served us well for some 15 years. A deciding factor in the purchase of the replacements is that many new transceivers have excellent coverage of 160. Hence, as the urge to acquire new equipment gets stronger, the repopulation of 160 increases at a rather steady rate.

Moving to the top band raises the question of what to do about an effective antenna. A conventional half-wave horizontal antenna, popular through the years on

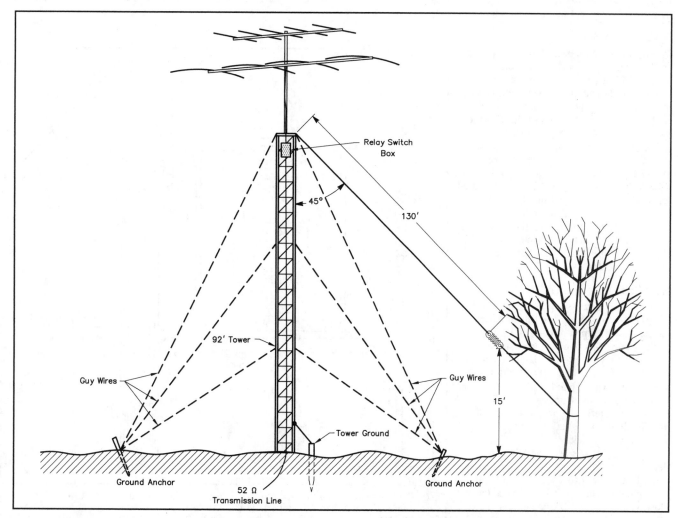

Fig 1—The W1CF sloper is arranged in this manner. Three monoband antennas atop the tower provide some capacitive loading.

this band, has limitations. Frequently, amateurs do not have room to put up 260 feet of wire. Moreover, the high angle of radiation from this type of antenna does not make it perform well as a DX chaser. What alternative then?

The writer, like many vintage DXers, is the proud possessor of a high, guyed steel tower festooned with monoband Yagis. I considered the several approaches to putting this combination to work without a major investment of time and money in order to have a *competitive* top-band signal.

My solution led to the construction of a quarter-wave sloper (also referred to as a half-sloper[1]) strung from the top of the tower and fed with 50-ohm coaxial cable connected through an existing six position coaxial switch. (See Fig 1.) The switch is remotely controlled from the operating position, allowing quick selection of other slopers which I plan to add. There is a directional effect produced by this type of sloping antenna which makes it desirable to have additional wires sloping in different directions.

Provided that an amateur has a tower, the incremental cost of adding the half sloper is negligible. A single antenna of this type involves just the purchase of two insulators and 130 feet (39.6 m) of copperclad wire. The relays for a system having more than one radiator would, of course, be an additional expense, but a modest one.

A 45-Degree Slant

The author's antenna slopes away from the tower in a southwesterly direction at an angle of 45 degrees to the tower. The bottom end of the radiator is fastened to a tree at a point 15 feet (4.6 m) above ground.

Rf is fed to the top end of the 130-foot wire by means of coaxial cable. From the end of the transmission line, rf is passed through one of the relay-operated coaxial switches to the antenna as shown in Fig. 2. A short length of copper wire connects the antenna to the center pin of one of the switches. The body of the relay enclosure is electrically grounded to the tower by means of the attachment plate and to the RG-17/U cable braid through the input cable connectors.

How Does It Perform?

The whole process of putting up a single antenna consumed one hour. But unlike many endeavors performed in haste in the middle of the winter, this one was very successful. Without taking time to trim the antenna, the full massive power of the TS-820 (90 watts key down!) was applied to the half sloper through a Bird wattmeter having a 250-watt element. The reflected power was less than one dial division over the 160-meter band. It virtually was unreadable.

With only two weeks of operating under my belt at the time of writing this article, I can give little more than a qualitative opinion on the operation of the antenna. However, in a recent 160-meter contest, my station seemed to be reasonably competitive, both on domestic and overseas contacts. I held a frequency for about an hour while chaining contacts using CQs and QRZ without being blown away by the competition. On overseas calls, the first or second try provided the wanted contacts. Most of the reports were RST 569.

Conversations after the contest with K0RF, who shares many of my antenna and operating thoughts, indicated that although my signals were down approximately 5 dB compared to K1PBW, who uses two top-loaded, quarter-wave radiators driven in quadrature (90 degree separation), my signals were well near the top of the New England pileup at his Colorado location. All this with a barefoot TS-820S—sigh!

The quarter-wave sloper "listens" well. I have heard K6SE (in the direction of the slope), and, surprisingly, PA0HIP and G3SZA with S9 signals. I do feel that this antenna is not the equal of a 1000-foot terminated Beverage receiving antenna, but it will bring in most of the multipliers that are on the air for the one night stand of an ARRL DX contest.

A 92-Foot Tower Helps

Inasmuch as this quarter-wave sloper appears to perform well, it is worthwhile to explore why. The W1CF 92-foot tower provides an advantageous height for putting out an attention-getting signal. Remember, the shield of the coaxial feed line is connected to the top of the tower. Although the ruggedness of the structure has little to do with radiating ability, I will mention in passing that it is made of heavy-duty galvanized steel. The bottom is bolted to a concrete base. Guy wires of $^3/_8$-inch (10-mm) cable without insulators are placed at three levels. The guy anchors are buried to a depth of 6 feet (1.8 m).

Atop the tower, three monobanders staggered at 5-foot (1.5-m) intervals on a 2-inch (51-mm) OD pipe act as capacitive loading. This large assembly must provide a fair amount of capacitance to ground, even at 160 meters. The "fat" tower and the parallel-connected guy wires provide relatively low inductance to ground.

Of course, each tower arrangement is found to be different. Several other amateurs using the quarter-wave sloper on the 40 and 80-meter bands have been enthusiastic about individual performances even though their masts were much shorter (50 feet). A detailed analysis of the actual antenna circuit is beyond my capabilities. What is apparent, however, is the following: (1) my particular tower provides a reasonably good ground return (counterpoise?), (2) the maximum-current point of the antenna is high, where it can do some good, (3) the measured feedpoint impedance is close to 50 ohms, (4) the match is

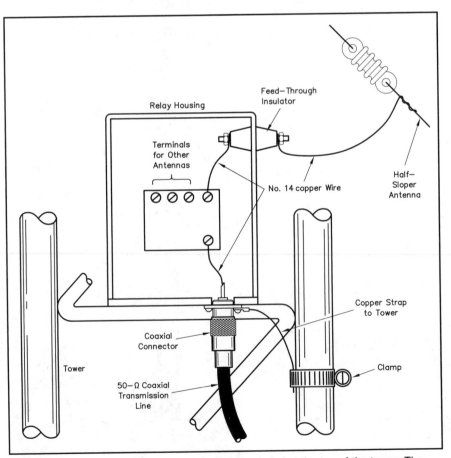

Fig 2—This drawing illustrates how the half-sloper is fed at the top of the tower. The use of remotely controlled relays permits the choice of two or more slopers in order to take advantage of the individual directional effects.

Relay Housing

Terminals for Other Antennas

Feed—Through Insulator

No. 14 copper Wire

Half—Sloper Antenna

Coaxial Connector

Tower

Copper Strap to Tower

Clamp

50—Ω Coaxial Transmission Line

relatively broadbandcd, and (5) the sloper provides an appreciable amount of vertically polarized radiation.

The writer feels that the quarter-wave sloper lends itself to many existing U.S. amateur installations with almost zero increase in cost. It seems to be reasonably competitive and is easy to erect. My guess is that an eager amateur with three spare positions on a relay-operated antenna

switching circuit could string three quarter-wave slopers from his tower at 60 degree intervals and probably obtain some worthwhile directivity at low radiation angles.[2]

The writer is indebted to Dr. James Lawson, W2PV, for helpful suggestions concerning this article. Words of appreciation also go to "Duke" Brown, W1ZA, for his installation assistance, and to Phil True,

W7AQB, for filling me in on his extensive experience on 75-meter phone with a similar installation.

Notes

[1] *The Radio Amateur's Handbook*, ARRL, 56th Edition, 1978.
[2] The quarter-wave sloper working against a good water-pipe ground suggests itself as being of use to a "cliff dweller" who wishes to drop a reasonably unobtrusive wire out a window of a high condominium or apartment.

Additional Notes on the Half-Sloper

My first rhetorical exposure to "half-sloper" antennas left me feeling that the person who lauded the concept belonged to some secret voodoo cult. The technique appeared to be laced with "black magic" with respect to the DX capabilities and simplicity of installation. At the time, I was entirely happy with my 40-meter "full-sloper" antenna, which had given superb DX performance over a three-year period. With change sometimes being good for the soul, I decided to look further into the matter. So during an "eyeball" QSO with Rush Drake, W7RM, I asked his opinion of the half-sloper antenna. He had used them on 80 and 160 meters with very good results. Not being of a mind to dispute a DX baron like Rush, I decided to "put up" (if I may resort to a pun), then "shut up" if need be.

The 40-meter full sloper was taken off the tower. The high end of the dipole was at 50 feet (15 m) and the low end was 7 feet (2 m) above ground. A 50-ohm coaxial feeder came off the center of the sloper at approximately 90 degrees. A TA-33 Jr. triband Yagi was located above the sloper, and a system of 16 buried radials (varied lengths of 60 to 110 feet—18 to 33.5 m) was fanned out below the sloper.

With all things remaining the same, exclusive of the 40-meter antenna just discussed, W1VD climbed my tower and "implanted" the new 40-meter half-sloper antenna. It had been cut to the traditional $L_{(feet)} = 234/f_{(MHz)}$. The shield braid of the coaxial cable was made common to the tower top near the driven-element insulator (Fig. 1). Then the feed line was taped to a tower leg at intervals all the way to the ground. It was then routed along the surface of the earth to a feedthrough panel which is used as an rf service entry to the shack. It should be mentioned that my purpose in having the buried radials has nothing to do with the 40-meter antenna. They were laid for use on 80 and 160 meters because the tower is employed as a vertical antenna (shunt fed) on those bands.

Antenna Adjustment

I had been told that it was a simple matter to adjust the half sloper for an SWR of 1. All that was supposed to be necessary was the pruning of the radiator length until an SWR of 1 was observed in the chosen

part of the band. I made my adjustments for 7025 kHz. It took nearly two hours of adding wire, removing wire and hoofing it into and out of the shack before the SWR bottomed out at 1.6:1. Bandwidth between the 2:1 SWR points was approximately 100 kHz. This was determined by readjusting the radiator for the lowest attainable SWR at 7100 kHz. In my installation, the radiator length was somewhat greater than $^1/_4$ wavelength. The best match was secured when the radiator was 3 feet (0.9 m) longer than the formula dictated. The enclosed angle between my unguyed tower and the half sloper is roughly 45 degrees. RG-8/U cable is used as the feeder.

Others who have worked with this type of antenna, but on 80 and 160 meters, tell of conflicting results with the radiator length. Two amateurs who erected 160 meter half slopers on 50-foot towers reported that the radiator had to be considerably shorter than $^1/_4$ wavelength and that an SWR of 1 was obtained. No doubt the reduced length can be related to the proximity of the wire to ground (added capacitance). Two amateurs who erected half slopers for 80 meters (on 100-foot or 30-m towers) said the lengths were precut to $^1/_4$ wavelength, and an SWR of 1 resulted. This suggests that each installation is unique, requiring some empirical work on behalf of the amateur. I hope to do some antenna scaling to 28 to 144 MHz soon. No doubt a model half sloper can be checked then for characteristic impedance, radiation pattern and radiation angle. For the present, anything I might claim would be pure conjecture.

As for performance, the 40-meter half sloper seems to greatly exceed the full sloper thus far. Even though it slopes off the west side of my tower, and supposedly has radiation reinforcement in that direction, I am receiving good reports from Europe to the northeast and South America to the south. This also was true of the full sloper, which tilted to the south. For the most part, my signal reports are 10 to 20 dB better than previously. This has been noted by three W8 stations in Michigan with whom I've maintained weekly schedules for the past two years. At 0100 UTC my 1-kW signal reports consistently run from 20 to 40 dB over S9 in Michigan, whereas they used to be on the order of S9 to 20 dB over S9. I have observed the same improvement with stations I contact fre-

quently in Texas and California.

Perhaps the major improvement in performance comes from the current portion of the antenna being raised to twice the original height, which is significant with any type of antenna. What role the tower plays in the overall system requires careful analysis. Perhaps such an investigation would dispel any black magic that seems to exist. But the half sloper does work, and mighty well.

One weak characteristic I noted is that when the upper insulator and feed connection point become covered with ice, the antenna is rendered useless. The SWR reads full scale in the forward and reflected directions, and the transmitter won't load into the system. A protective covering is suggested for that part of the system if you live where sleet storms are likely to occur.

W7RM suggested a unique way to employ half slopers. Two or four of them are placed on the tower. Opposite wires can be joined to the feeder by means of a remote relay to convert any two half slopers to an inverted-V antenna. This gives the operator a choice between low angle radiation with the half sloper and higher angles of radiation with the inverted V. Four half slopers can be installed 90 degrees apart on the tower, then switched for any one of four chosen points of directivity. A remote switch would be used for this also. — *Doug DeMaw, W1FB*

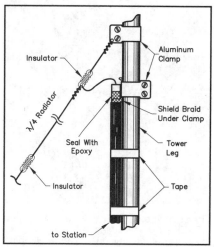

Fig 1—A method of installing and feeding a half-sloper antenna.

A 75-Meter DX Antenna

The long, cold winter nights in VE6 land, where I live, would seem like ideal times to work DX. At this time of year, though, the 20-meter band is as silent as a powerless transmitter. One, therefore, must use the lower bands to pursue any DX activity during the hours of darkness. I chose to operate in the 75-meter band using an inverted-V antenna. Reaching stations from coast to coast presented no problem, but all those DX stations I heard failed to respond to my calls. A change of antenna seemed desirable.

With a form of half sloper that I rigged up the DX barrier was broken. XE2AX became my first DX contact, followed two days later by a QSO with ZL2BT who radioed back a gratifying 5-9 report. That initial success led to the expansion of the sloper system to include two other half slopers as shown in the drawing. Radiation from a sloping antenna tends to be maximum in the direction of the slope. For that reason, with the antennas spaced 120 degrees apart, I'm able to take advantage of the directional characteristics of the system that offers three different signal patterns. Each of the three wires slopes down-

This utility box, which is mounted atop W7AEK's tower, houses two relays that are used to select any one of three sloping radiators connected to the terminals at the back of the box.

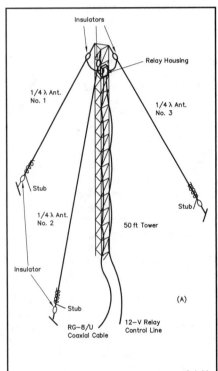

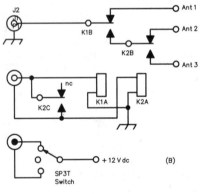

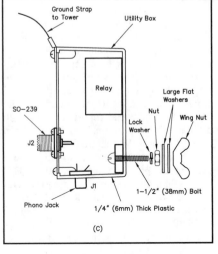

ward at a 45-degree angle from the apex of the tower.

Feed points for the radiators are at the top of the tower where two relays enclosed in a plastic utility box (see photo) provide the necessary switching for selecting any one of the three antennas. This arrangement eliminates the need for separate transmission lines. An accompanying diagram illustrates the relay circuit. Each radiator is cut to approximately 1/4 wavelength at the operating frequency. Length, after final adjustment, should be between 60 and 65 feet, depending upon the selected frequency and the conditions at the antenna site. When trimming such an antenna for the 75- or 80-meter bands, tuning stubs, such as shown in the drawing, simplify the work. These may be shortened or lengthened as needed. A change in length will produce a resonant frequency difference of about 0.1 MHz per foot.

The relays I chose have 10-ampere contacts and operate from a 12-volt control circuit. The plastic keys on the octal bases of the relays had to be trimmed in order for the units to fit inside a $1^{1}/_{2} \times 5 \times 2^{1}/_{2}$-inch (38 × 127 × 64 mm) Radio Shack utility enclosure. Pin-to-pin leads are short. Wiring is done with no. 14 hookup wire. The outer conductor of the coaxial cable is common to the tower as shown in the illustration.

Reports that resulted from the first week of testing indicated a 25-dB change as the selection of radiators was rotated. During the last weekend of the ARRL DX Contest, l logged the following prefixes on 75-meter ssb: VP1, VP2, XF1, XE1, KH6, KL7, PJ8, KP4, TG9, ZL, HD and KZ. I'm indeed pleased by DXing now with my $30 half-sloper system.—*S. Timothy Hopps, W7AEX/VE6, ex-W1WZR*

The W7AEK 75/80-meter DX antenna system is illustrated above. For directional effect three individually operated radiators are used. The circuit for the relay-operated antenna selector switch is at B. The utility box housing the relays is shown in a cutaway view at C. lengthening or shortening the radiators is simplified by the use of the tuning stubs in Drawing A. Note that the shell of J1 is not grounded.

J1—Phono jack.
J2—SO-239 chassis-mounted coaxial connector.
K1—Same as K2 but used as an spdt relay.
K2—Dpdt 12-V dc relay, 10-A contacts, Radio Shack no. 275-208 or equiv.
P1—Phono plug.
Utility box—Radio Shack no. 270-233 or equiv.

From *QST*, February 1991 (Technical Correspondence)

More On The Half Sloper

"The Half Sloper—Successful Deployment is an Enigma," was a popular article.[1] Many amateurs had success with the antenna, others did not. Why? I found that if the half sloper was fed against an empty tower (a tower supporting no other antennas above the sloper), the sloper could not be made resonant at the desired frequency (where the length of the sloping wire was $1/4 \lambda$ long). Measured on an antenna range, the pattern of such a half sloper is more or less omnidirectional. Those hams who had successfully resonated the half sloper had suspended the sloping wire from a tower that supported a "plumber's delight" Yagi antenna. Those amateurs reported that the half sloper had directional characteristics and was a good DX antenna.[2]

Fig 1A shows a half sloper dimensioned for the 75-meter band. This configuration is similar to the version I originally modeled. The input impedance (Z_a) for this antenna at 4 MHz (as calculated by the ELNEC version[3,4] of *MININEC-3*) is $374 + j1109 \, \Omega$, which shows clearly the antenna is not resonant. According to *ELNEC*, the antenna is resonant at a much lower frequency (2.714 MHz). At this frequency, the antenna's impedance is $22 \, \Omega$. The ratio between the actual resonant frequency and the desired resonant frequency is 0.68 (2.714 ÷ 4). The curves of Fig 2 show radiation patterns for this redimensioned antenna over average ground ($\sigma = 3$ mS/m, $\varepsilon = 13$). The azimuthal pattern is more or less omnidirectional, which again is in agreement with the experimental measurements, and the predicted gain is rather low.

Fig 1B shows a half sloper suspended from a 50-foot tower that supports a 20-meter 3-element Yagi. The Yagi is a monoband, wide-spaced beam (the spacing between the reflector and driven element is $1/4 \lambda$; $1/8 \lambda$ between the driven element and director). The input impedance for this half-sloper system at 4 MHz is $33 - j104 \, \Omega$. Certainly, this is a more reasonable impedance, and the trend agrees with expectation. Lengthening the sloping wire from 58.43 feet to 68.27 feet (0.274 λ) resonated the antenna at 3.95 MHz (input impedance $46 \, \Omega$). The curves of Fig 3 show radiation patterns for this antenna. Clearly, this half sloper has directivity (F/B of 8 dB at 25°) and reasonable gain. The 20-meter Yagi is an important part of the antenna system. It plays the part of the disc of a "skeleton discone" antenna. This analogy is clearer if you examine the current in the vicinity of the feed point. The amplitude of the currents on the

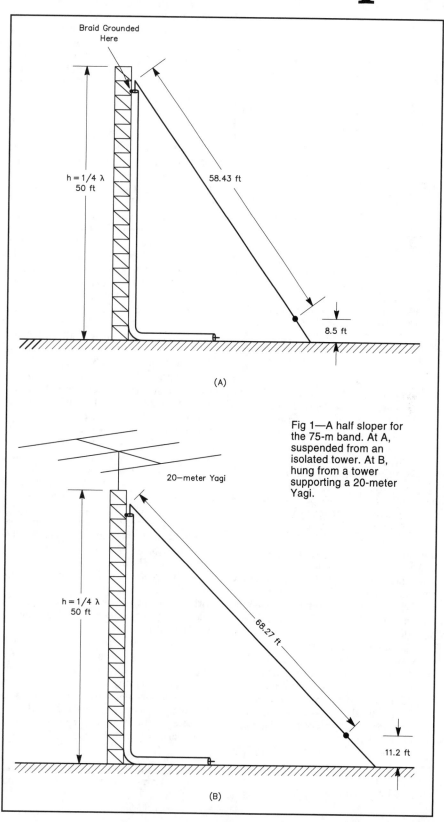

Fig 1—A half sloper for the 75-m band. At A, suspended from an isolated tower. At B, hung from a tower supporting a 20-meter Yagi.

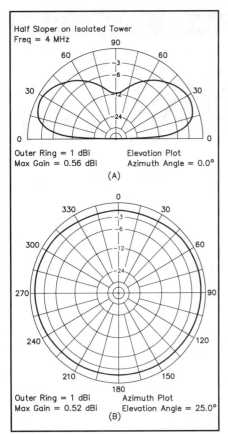

Fig 2—Vertical-plane pattern (A) in the plane containing the antenna, and azimuthal pattern (B) for the antenna of Fig 1A, calculated by *ELNEC* for an antenna over average ground (f_o = 4 MHz).

fed end of the sloping wire and on the pipe mast that supports the Yagi are more or less equal and approximately in phase. The amplitude of the current on the tower is much smaller (by a factor of three) with a large phase difference (about 135°).

The antenna-input impedance, the length of the sloping wire required for resonance and the antenna pattern all depend on the tower height, the angle between the sloper and tower, the type of Yagi on the tower and the direction in which the Yagi is pointing. For this study, the Yagi was aimed in the direction of the sloping wire—toward 90° azimuth. If the Yagi is 90° to the wire, the gain at low angles is essentially the same, but there is a small change in the shape of the radiation pattern (about 0.25 dB increase in the field at high elevation angles), and a small change in the input impedance.

After convincing myself that *ELNEC* could accurately model the antenna system—the sloping wire, tower with Yagi atop and guys (if one wished to include guys in the model)—I made a detailed study of half slopers[5] and reached the following conclusions:

1) For an 80-meter half sloper, the pattern and gain are not strongly dependent on the type of Yagi on the tower. However, the sloper's impedance (particularly the reac-

tance, and hence the length of wire required for resonance) depends on the size of the Yagi. The larger the Yagi, the more effective it is as a ground plane, and the more nearly the length of the sloper for resonance approaches that of a ground plane 1/4-λ monopole. For 160-meter operation (with the half sloper suspended from a 95-foot tower), a 20-meter Yagi is too small to be an effective ground plane. According to *ELNEC*, a resonant length for the half sloper can be found, but this length is much longer than 1/4 λ: 203.4 feet or 0.39 λ, and the input impedance is rather high for coax feed—about 400 Ω instead of about 50 Ω. A 3-element, 40-meter Yagi on the tower, however, is just fine. The input impedance (according to *ELNEC*) is 55 Ω and the sloping-wire length is 145 feet for resonance at 1.9 MHz (length 0.28 λ).

2) For a 50-foot tower, there isn't much choice in regards to the angle that an 80-meter half sloper makes with the tower. For a 95-foot tower, however, this angle can be varied. As this angle increases from 25° to 50°, the azimuthal pattern becomes more and more cardioidal, the low-angle radiation decreases marginally (by about 0.3 dB for the range of angles studied), and the high-angle radiation increases (by about 1.8 dB). The antenna's resistance decreases as the angle between the sloping wire and the tower decreases: For an 80-meter half sloper attached to a 95-foot tower, the resistance changes from 21 Ω to 13 Ω when the angle decreases from 50° to 25°.

3) Although a 40-meter half sloper (attached to a 50-foot tower with a 3-element 20-meter Yagi atop) can be resonated (resonant length is about 0.24 λ), the radiation patterns no longer resemble those that characterize 80- and 160-meter half slopers. In fact, the differences are astonishing. The pattern and impedance depend very strongly on the azimuthal position of the Yagi. According to *ELNEC*, the antenna resistance (for antennas dimensioned for resonance), changes from 76 Ω to 127 Ω when the Yagi is rotated from the position where its elements are parallel to the plane containing the sloping wire and the tower to a position perpendicular to this plane. Antenna lengths for resonance are 32.37 feet and 33.88 feet, respectively. The azimuthal pattern (which is roughly bidirectional) is maximum, but skewed in the plane containing the sloping wire and the tower, and in the plane broadside to this plane for the described azimuthal positions of the Yagi.

I knew the answer to the enigma in regard to successful deployment of the half sloper a long time ago, but now—thanks to modern antenna design using a computer— I have a better understanding of the characteristics of this antenna. And we now have the ability to generate radiation patterns for particular configurations.—*John S. Belrose, VE2CV*

Notes

[1] J. Belrose, "The Half Sloper—Successful Deployment is an Enigma," *QST*, May 1980, pp 31-33.

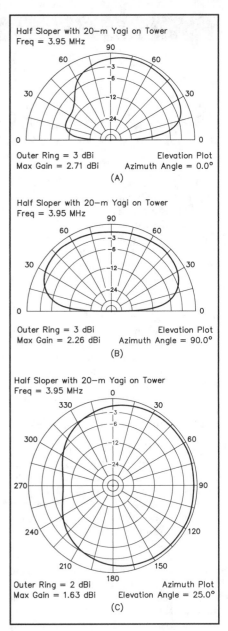

Fig 3—Vertical-plane patterns: A, in the plane containing the antenna (90° azimuth corresponds to the direction of the drooping wire away from the feed); B, in the plane orthogonal to the antenna; C, azimuthal pattern for the antenna of Fig 1B over average ground, as calculated by *ELNEC* (f_o = 3.95 MHz).

[2] J Belrose, "An Update on Sloping-Wire Antennas," *QST*, Sep 1984, p 40.

[3] *ELNEC* is available from Roy Lewallen, W7EL.

[4] *ELNEC* and *MININEC* do not accurately predict the real gain for antennas close to the ground particularly for horizontal antennas close to poor ground (height less than 0.2 λ). However, for the present analysis, the gains predicted by *ELNEC* are thought to be about right, particularly for the tower and Yagi arrangement, since there is little current on the tower.

[5] I have a folder full of patterns for different tower heights (60 and 100 feet), different types of Yagis (monoband 3-element, 20- and 40-meter Yagis), and 5-element triband Yagis for 20/15/10 and 40/20/15 meters.

By Don Kirk, WD8DSB

From *QST*, March 1998

A Reduced-Size Half Sloper For 160 Meters

Here's a limited-space antenna that will put you on the Top Band!

Have you ever thought of trying 160 meters? Wintertime is the perfect time for exploring the "Top Band." The atmosphere is relatively quiet, so you can mine 160 meters for all it's worth—lots of stateside contacts and plenty of DX!

If you live on a small lot, you're probably muttering, "This guy must be kidding!" Yes, most 160-meter antennas are *big*. A full-sized half-wavelength dipole is about 260 feet long! But with a little creative design work, it's possible to build limited-space antennas that will get you on the band without sacrificing too much real estate. They aren't as efficient as the full-sized variety, but they *will* get you on the air. Isn't that what counts?

My Approach

Maybe you've seen designs for half-sloper antennas. As the name suggests, these antennas literally slope down to the ground from a tree or tower. Many 160-meter half-sloper designs require supports that are at least 50 feet tall. The problem is that my tower is only 40 feet tall. And what about hams (like you, perhaps) who don't own towers at all? It was time for a different solution.

If you want maximum efficiency from the antenna, you'll need to install some radials.

The typical half-sloper design uses the tower as one half of the antenna by connecting one side of the feed line (normally the shield of the coax) directly to the tower. The center conductor of the coax connects to a $1/4$-wavelength wire that slopes back down toward the ground. My idea was to design a half-size ($1/8$ wavelength) sloper using inductive loading. I decided to place the coil directly at the top of the wire (see Figure 1). Placing the coil at this location requires the least amount of inductance, but it's a trade-off with efficiency.

Those hams not blessed with a 40-foot tower can substitute a convenient tree or other support (see Figure 2). Just use a wire running vertically to ground as your substitute "tower." More about this in a moment.

Construction

A 26-inch length of $3/4$-inch PVC pipe (outside diameter of 1.05 inches) is used as the coil form. I chose a 26-inch long pipe because I wanted to wind the coil on one end and attach the opposite end to my tower. You can certainly use shorter pipes

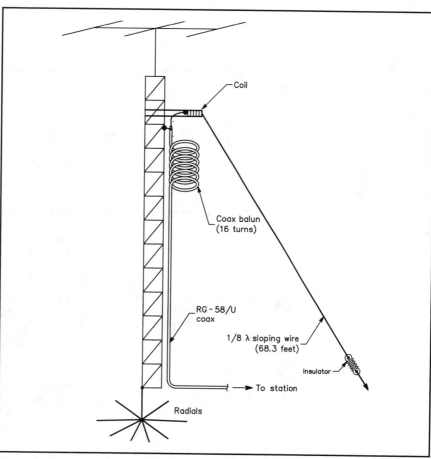

Figure 1—Even a 40-foot tower makes a convenient support for a 160-meter half-sloper antenna.

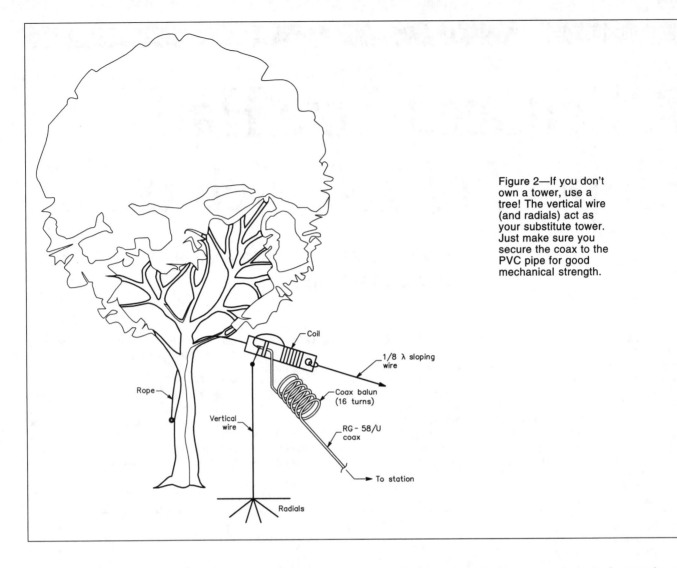

Figure 2—If you don't own a tower, use a tree! The vertical wire (and radials) act as your substitute tower. Just make sure you secure the coax to the PVC pipe for good mechanical strength.

according to your particular requirements. Bear in mind that this coil is designed for use with 100-W transceivers.

Wind 90 turns of 16-gauge enamel-coated wire at one end of the pipe. Keep it tightly spaced to produce a 4½-inch long coil (see Figure 3). Wrap the coil with two layers of electrical tape (Figure 4). Solder the center conductor of your coax to one end of the coil as shown. Apply silicone caulk to the exposed ends of the coax. Solder a

Yes, most 160-meter antennas are big. A full-sized half-wavelength dipole is about 260 feet long!

68.3-foot piece of wire to the opposite end (this is the sloping portion of the antenna).

Install the coil on your tower and clamp the coax shield to the tower leg using a stainless-steel hose clamp (Figure 5). Make an RF choke (balun) by coiling 16 turns of your coax into an 8-inch diameter coil. You can use electrical tape or tie wraps to hold it together. Tape the choke to the tower leg 2 feet below the feed point. Slope the ⅛-wavelength wire back to terra firma and ter-

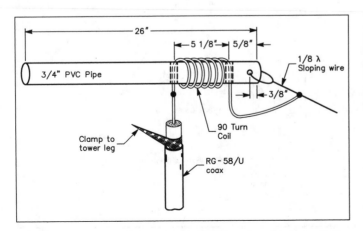

Figure 3—Construction details for the PVC mounting pipe and coil. For a tree installation, secure the coax to the PVC pipe behind the coil.

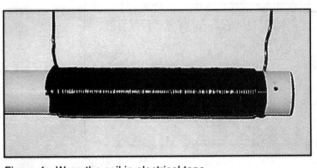

Figure 4—Wrap the coil in electrical tape.

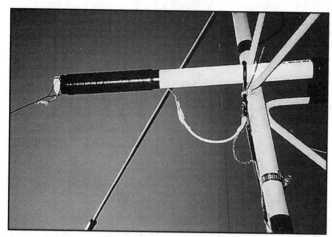

Figure 5—For a tower installation, attach the PVC mounting pipe horizontally and snake the coax down the leg of the tower.

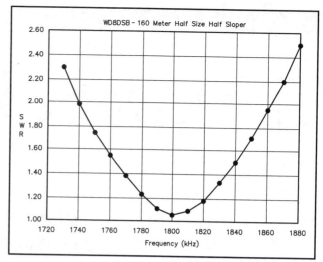

Figure 6—This is the SWR curve for my installation which uses an 8-foot ground rod and no radials. Notice that it is resonant at the bottom of the band. To make the antenna resonant higher up the band, just trim the length of the sloping wire.

minate it with an egg insulator about 7 feet above the ground to avoid contact with people and pets.

If you're taking the "tree tower" approach, install the coil as high as possible (preferably 40 feet or higher). With some clever rope-and-slingshot manipulation you can haul the coil into the desired position without actually climbing into the tree. Before you take the coil skyward, use some electrical tape to secure the coax to the PVC pipe behind the coil. If you allow the coax to simply dangle from the center-conductor solder connection, it will break in short order! Make the same RF choke as described in the tower installation and attach it to a nearby branch, or just let it dangle. Finally, solder the coax shield to a 40 or 50-foot length of wire. When you have the coil assembly at the proper height, bring this wire down to the ground as vertically as possible. Make sure it doesn't come in contact with the tree trunk or branches.

I manage to get away with using an 8-foot ground instead of the radials shown in Figures 1 and 2. But if you want maximum efficiency from the antenna, you'll need to install some radials. The radial wires attach to the base of the tower or, in the case of a tree installation, to the bottom of your ver-

tical wire. Don't worry about the lengths of these radials. Just run as many as you can for as long as you can. You can bury the radial wires, or simply lay them on top of the soil.

Results

My antenna loaded great and worked superbly right from the start without an an-

My idea was to design a half-size sloper using inductive loading.

tenna tuner. The SWR was plotted (Figure 6) and the 2:1 SWR bandwidth was 120 kHz. You'll note that my antenna is resonant at the bottom of the band (near 1.8 MHz). If you want it to resonant higher in the band, just trim the sloper wire. If you opt for tree mounting, you may have to tweak and trim the antenna a bit to compensate for any detuning caused by the tree itself. You can also modify this design for use on other bands (80 meters, for example) with the inductive reactance of the coil being 500 Ω for the desired band.

If you're squeezed for space, give this antenna a try. Of course, you can also modify the design for use on other bands (80 meters, for example). Have fun!

By Richard A. Lodwig, W2KK, ex-W2IHJ, ex-W3GNK, ex-K2ODT   From *QST*, April 1977

The Inverted-L Antenna

You can't put up a low-band half-wave dipole for local work because you're cramped for space? Unable to lay out a good radial system for a vertical antenna to work DX? This antenna may be just what you're looking for—it has the performance advantages of both, without the disadvantages.

Amateur Radio operation in the low hf bands has historically been a challenge, especially for those persons without large amounts of antenna real estate available. In order to operate on the 160, 80/75, and 40-meter bands in locations where either or both minimum space and few vertical supports are available, antenna compromises are generally required. Such compromises generally have a significant effect on the bandwidth, efficiency and direction of radiation of the signals. Many antenna types have been proposed and constructed which try to "optimize" the variables involved.

This paper introduces a new type of antenna, which was developed and tested, including comparison tests against several other popular antenna types. The antenna has exhibited excellent performance characteristics during these tests and offers some interesting installation advantages, such as no requirements for ground radials and short physical span between supports: In addition, it has been intentionally designed to provide simultaneous low-angle radiation (vertical polarization) for DX work and high-angle radiation (horizontal polarization) for short distance and local work via sky-wave signals. This article discusses some of the performance factors of this inverted-L antenna, as well as explains how and why it and some of the other popular 160, 80, and 40-meter antennas operate in a practical environment.

On the 40-meter band, the daytime D-layer absorption is even lower than on 160 or 80 meters, thus enabling communication up to about 1,000 miles during the day. Up to about 200 miles, the propagation is via high-angle vertically or horizontally polarized sky wave; from 200 to about 1,000 miles, moderate to low-angle vertically or horizontally polarized sky-wave signals provide the communications path. During the evenings, communication on 40 meters is possible from about 500 miles to as far as the limits of darkness permit; the signals are propagated as low-angle horizontally or vertically polarized sky waves. Communication over distances closer than 500 to 1,000 miles is usually impossible on 40 meters during the evening, because the high angle signals for this distance (within the "skip distance") penetrate the normally reflective F-layer.

At this point, let us examine a principle which relates the propagation mode to the relative performance of several antenna types: The effectiveness of any efficient antenna system is dependent upon how well the particular system (including the surrounding environment) couples energy into an efficient mode of propagation between communicating stations.

The convenience of a single antenna usable for both local and DX work is noteworthy.

Note that this principle implies several necessary conditions for maximizing the communicating efficiency: The antenna itself must be efficient (low losses, little energy dissipated by coupling to nearby objects). One or more efficient propagation modes must exist between the communicating stations since no antenna is capable of enabling communication when there is no suitable propagation mode. The antenna must couple energy efficiently into at least one of the propagating modes.

Taking all of these considerations into account, some general conclusions concerning the desirable properties of antennas for the 160, 80/75, and 40-meter bands can be drawn.

1) Close-in daytime communication would be optimized by an antenna which provides some degree of high-angle radiation (about 60-degrees elevation angle on 160 and 80 meters, and 30 degrees on 40 meters). Either horizontal or vertical polarization is suitable at these angles.

2) Close-in nighttime communication (where skip permits) would be optimized by an antenna which provides some degree of high-angle radiation (40 to 73 degrees) on 160 or 80 meters with either horizontal or vertical polarization suitable.

3) Distant nighttime communication on any of the three bands requires low-angle radiation (40 degrees or less) in either polarization. Generally speaking, the lower the elevation angle that an antenna system provides (closer to the horizon), the better will be the communication at extreme distances.

It is obvious that some of the most popular antenna systems used by amateurs provide efficient coupling to either high-angle or low-angle propagation modes, but not both simultaneously. This results in a compromise, whereby good results may be obtained for close-in daytime and nighttime communications, but not for distance or vice-versa. But there are certain antenna types which can be designed to provide efficient operation and coupling to both the close-in and distant propagation modes.

Simultaneous High and Low-Angle Radiation

The configuration for the inverted-L antenna was arrived at through a process which asked the question, "Is there an antenna type possible which will provide high-angle coverage (for local and moderate distance communication) and low-angle coverage (preferably down to the horizon for DX communication) simultaneously in a compact configuration which is reasonably independent of ground loss (thereby obviating the need for an elaborate ground radial system)?" After much thought, a configuration was found which theoretically met these requirements. An 80/75-meter and a 40-meter version of the antenna were built and the 40-meter model was tested directly against other antennas in a set of "blind" experiments. The results of the experiments confirmed that the radiation properties of the antenna were noticeably better than the four basic antenna types used in the "average" type of amateur operation involving both local and DX communication on 160, 80/75, and 40 meters, namely a horizontal dipole, a quarter-wave vertical, a half-wave vertical, and an inverted V.

Fig. 1 illustrates the basic inverted-L configuration. The antenna consists of a quarter-wavelength vertical element and a quarter-wavelength horizontal element which are joined at the central coaxial feed point. The feed line can be run off at an angle away from the plane of the antenna (as in Fig. 1), in the plane of the antenna, or in a direction perpendicular to both legs. The height of the feed point should be at least a quarter-wavelength above ground (but can be higher if feasible) in order that the vertical radiating wire does not reach the ground. (A later section will discuss variations of the basic antenna to allow for installation of the antenna at heights lower than a quarter wavelength above ground.) In the basic configuration, the antenna can

be considered to be either a horizontal dipole with one of the arms bent downward by 90° until it is vertical, or a half-wave vertical dipole with its top arm bent 90° until it is horizontal, or a 90° inverted-V antenna which is rotated in its plane by 45°.

Note several desirable factors which such a configuration results in. The vertical arm of the inverted L is a quarter-wave element above ground, but with the high-current end of the dipole located at the top of the vertical structure (not the bottom, as with a quarter-wave monopole). This results in relative independence from ground-loss coupling (since only the high-voltage, nonradiating end of the antenna is near the ground) and good low-angle vertically polarized radiation (since the active high-current radiating portion of the antenna is at a substantial height above ground).

The VSWR bandwidth of the inverted L is somewhat broader than that of an inverted V and slightly less than that of a dipole.

The horizontally polarized section of the antenna, being located at least a quarter-wavelength above the ground, provides good high-angle radiation in a manner similar to that of a full-sized horizontal dipole.

Naturally, the amount of signal radiated by the inverted-L antenna in low-angle vertical polarization is not as strong as that of a full half-wave vertical antenna, since we

are only providing one of the two arms of the antenna. Similarly, the amount of high-angle signal radiated by the inverted -L is not as strong as that provided by a full half-wave horizontal dipole antenna (again, because we are only providing one arm of the two which the dipole normally has). However, the inverted-L antenna provides a substantial amount of both types of radiation, and thus provides better "compromise" performance for the standard types of amateur communication which involves a mixture of high and low-angle radiation requirements.

The approximate radiation patterns of the inverted-L antenna in the E-W and N-S direction are shown in Fig. 2. Note the relatively uniform amount of total radiated power in both the N-S and E-W direction, independent of the elevation angle. These are theoretical patterns, calculated for perfectly conducting ground, but calculations for an imperfect ground show that the patterns are relatively independent of the electrical properties of the earth. In directions other than the E-W or N-S direction, there is always both a substantial amount of low-angle vertical polarization and high-angle vertical or horizontal polarization. It can be seen by examination of these radiation patterns that the antenna should provide effective local and DX coverage for any of the three low frequency bands.

With a 50-Ω coaxial feed line the VSWR bandwidth of the inverted L is somewhat broader than that of an inverted V and slightly less than that of a dipole, primarily because of the stronger coupling between the adjacent arms of the antenna. About 180 kHz of the 160-meter band or 375 kHz of the 80/75-meter band can be covered with a VSWR of less than 3:1, and the full 40-meter band can be covered with a VSWR of less than 1.8:1. (Later, it will be shown how a short "tail" can be added to the vertical element to permit easy adjustment of the exact center frequency to any desired spot for 160 or 80/75 meter versions of the antenna.)

Theoretical and Experimental Comparison of Types

An experimental program was undertaken to try to verify the predicted radiation performance of the inverted L. Forty meters was chosen to be the frequency band for the test, since there are useful and stable local and distant foreign broadcast signals which could be used for signal strength comparisons. Use of a-m stations facilitated the signal strength measurements by enabling measurement of the received carrier level.

A horizontal half-wave dipole was erected as a reference antenna. Tested against this reference antenna were a quarter-wave vertical, a half-wave vertical, an inverted-V, and the inverted-L antenna. The method for measurement of the antenna performance was to select a local or distant a-m carrier for reception by a calibrated receiver, with the two antennas under test

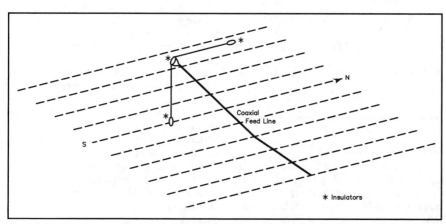

Fig 1—The inverted-L antenna. As an aid to discussing radiation patterns, the antenna is shown with the horizontal portion running north and south. The vertical and horizontal sections are each a quarter wave in length for the frequency band of operation.

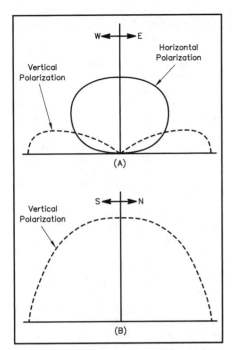

Fig 2—Calculated radiation patterns of the inverted-L antenna having the orientation shown in Fig 1. These patterns show performance for a perfectly conducting earth, but calculations for imperfect ground indicate that the patterns are relatively independent of the electrical properties of the earth.

being fed to an spdt rf switch. The peak carrier level was recorded during a period of about 30 seconds for each of the two alternate antennas and was tabulated. The test was performed in a "blind" fashion with a noninvolved party performing the connection of the two antenna connectors to the spdt switch prior to recording of data. Only after completion of the test were the antennas identified. Several hours of measurements were performed at different times during the day and night for local and distant signals. All of the results (for signals in various directions) were averaged in order to obtain the "average" effectiveness of each candidate antenna compared to the dipole standard. The results of the evaluations are summarized in Table 1.

Of particular significance is the performance of the inverted L with respect to all other antenna types (exceeded in performance by only the half-wave horizontal dipole for local contacts and the half-wave vertical dipole for DX signals). Note the effect of the radials on the performance of the quarter-wave monopole. The radials were physically removed for the "no radial" test, and only a minor VSWR effect was noted when comparing data before vs. after radial removal.

It should be noted that during the measurement series the general trends of the received signal measurements usually substantiated the average trends presented in Table 1. Of course, particular measurements

Table 1
Received Signal Level from 40-Meter Test Series

ANTENNA	LOCAL (<1000 MILES)	DISTANT (>2000 MILES)
Horizontal half-wave dipole λ/4 above ground	Reference	Reference
Quarter-wave monopole, 12 λ/4 radials	–8 dB	+2 dB
No radials	–10 dB	–4 dB
Half-wave vertical dipole	–15 dB	+9 dB
Inverted V, 0.2-λ vertex ht.	–3 dB	+3 dB
Inverted L	–2 dB	+6 dB

taken at a particular time with a particular signal may momentarily show signal strengths which are different than that indicated in the table.

The relative effectiveness of the inverted-L antenna compared with the others evaluated for the average amateur operation is clear from the table. For local operation, the inverted L was not as effective as the horizontal dipole (2 dB worse), but it is slightly better than the inverted V (about 1 dB better) and significantly better than the quarter-wave monopole and half-wave vertical dipole. When used in providing long-distance communications, the inverted L is not as effective as the half-wave vertical dipole (3 dB worse), but is significantly better than the inverted V (by 3 dB), horizontal half-wave dipole (by 6 dB), and quarter-wave monopole (by 4 dB, even when the monopole had a radial system).

The antenna has exhibited excellent performance characteristics during these tests.

Variations on Basic Inverted L

For the basic inverted L, the approximate length of each leg (in feet) can be calculated by dividing the desired resonant frequency (in MHz) into 230. If it is desired to provide for a degree of adjustment of the resonant frequency (especially in the

160 and 80/75-meter versions of the antenna), an additional length of wire may be added at the accessible bottom of the vertical section. This additional wire can be either vertical or horizontal since hardly any radiation occurs at this point in the antenna. The tuning wire should not be a portion of the main antenna structure, so that its attachment can be accomplished rapidly. An alligator clip or other similar means of attachment is quite acceptable, since very little rf current flows across the connection. These tuning wires can be preconstructed, such that with no wire added the antenna is resonant near the high frequency end of a particular band. Adding a particular wire will lower the resonant frequency to a preset new frequency lower in the band.

If the geometry of a particular installation does not permit equal lengths for the vertical and horizontal portions of the antenna, the legs may be made of unequal lengths (within reason) as long as the overall antenna length is 460 divided by the desired resonant frequency. As an alternative, the leg lengths may be kept equal and the feed point located at a point other than the point at which the antenna bends by 90°. For most effective operation of the antenna, the feed point should be located at the junction of the horizontal and vertical sections of the antenna.

Electrically loaded (shortened) antennas, which are commercially available from a number of sources, or continuously loaded types (such as Slinky dipoles) can be employed where space does not permit installation of a full-sized antenna. Even trap dipole antennas can be used to provide multiband coverage in a single inverted L antenna structure. Tuning of the antenna at the bottom of the vertical section (similar to the full-sized version) is practical for these alternative configurations.

The inverted-L antenna has been in use at the author's station since the spring of 1974. Performance of the antenna has been quite satisfactory, to the point where other antennas (including horizontal and vertical dipoles) have been removed and are no longer used. The particular advantage of the antenna (excellent local and DX coverage in a single antenna) is easily observed when operating with this antenna. The convenience of a single antenna usable for both local and DX work is noteworthy.

It should be mentioned that I am not aware of any previous disclosure of this particular antenna structure, but I am certain that it must be in use by others who have installations which dictated the particular horizontal-vertical configuration of the invertical L. It is not my intention to claim "discovery" of this antenna, nor to advocate its use in preference to other antenna types. The purpose of the article has been to enable the reader to gain a little more understanding of how various antenna types operate, especially the concept of effective antenna coupling to propagating modes in the 160, 80/75 and 40-meter bands.

By Dennis Monticelli, AE6C

From *QST*, July 1991

A Simple, Effective Dual-Band Inverted-L Antenna

Classic ham-antenna types are classics for a reason: They've worked predictably and well for several generations of hams. Here's how to put such an antenna—the inverted L—to work on the low bands.

B uilding antennas is fun! These days, we purchase and operate sophisticated transceivers, but we still build many of our aerials. It satisfies us to create something that really "gets out." If only we each had several acres of tall trees. . . sigh. More often than not, though, we find ourselves crowded onto small city lots among neighbors who don't share our love for antennas. This presents quite a challenge, especially to low-band enthusiasts. What to do?

Build an inverted-L antenna! An 80-meter inverted L takes up only a quarter-wavelength of horizontal space (64 feet) and can be built to blend into the background. You'll find it satisfying and it will work well on 40 meters, too. Inverted-L antennas are also easy on the pocketbook and use commonly available materials.

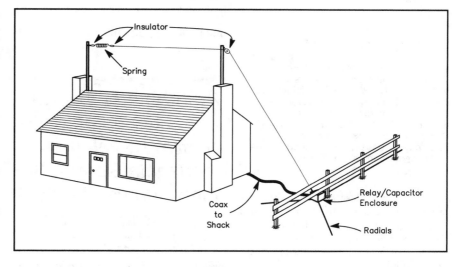

Background

Inverted-L antennas aren't new; they've been around for a long time. Most often they're used on 160 and 80 meters as an alternative to a $^1/_4$-λ vertical where support height is scarce. They're proportioned according to this single guideline: make as much of the antenna as possible vertical and bend the rest over. Usually, $^1/_4$-λ inverted Ls end up roughly $^1/_8$ λ vertical and $^1/_8$ λ horizontal (Fig 1A).

Fig 1B shows a variation on the basic inverted L. With its total wire length of $^3/_8$ λ, this version offers several advantages over the quarter-wavelength configuration. Most importantly, the $^3/_8$-λ antenna's radiation resistance and feed-point impedance are higher, which decreases the effect of ground losses.[1]

Because this antenna's maximum-current point occurs $^1/_8$ λ above ground level (instead of at ground level, like the $^1/_4$ λ version), the antenna can also "see out" beyond obstacles better than the shorter version. Computer simulations for a $^3/_8$-λ 80-meter inverted L[2] reveal an essentially omnidirectional pattern with slight directivity (1.5 dB) in line with the horizontal wire and pointing away from the far end.[3] The takeoff angle is favorable for short and medium-haul contacts (out to a few thousand miles).

> *One of the nice things about inverted Ls is that they fit easily on most city lots.*

A Free Band

What makes the $^3/_8$-λ inverted L especially interesting is that it gives you decent performance at twice the fundamental frequency (that is, where the total antenna length is $^3/_4$ λ—Fig 1C). You can build the antenna for 160 and 80 meters, 80 and 40 meters, and so forth. At twice the antenna's fundamental operating frequency, the antenna is resonant and has two current maxima: one at the feed point and another in the middle of the horizontal wire. The antenna acts essentially like a quarter-wave vertical end-feeding a horizontal half-wave dipole. The radiation pattern looks much like that of a dipole because the horizontal wire dominates the antenna's overall performance at moderate to high elevation angles, but the vertical wire contributes valuable low-angle radiation. At low elevation angles, the antenna has little directivity; at higher angles,

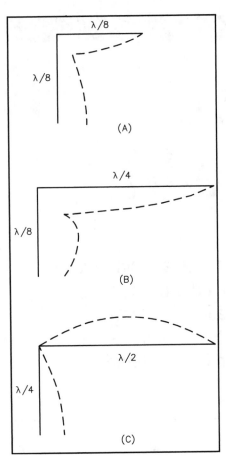

Fig 1—At A, the basic inverted L commonly used on the lower-frequency ham bands. The dotted line represents current distribution. The $^3/_8$-wavelength inverted L shown at B features a more favorable current distribution. At twice the fundamental (C), the antenna at B acts as a $^3/_4$-λ wire. Note the two current maxima. The antenna behaves like a quarter-wave vertical end-feeding a half-wave dipole.

the pattern more closely resembles the pattern of a dipole at the same height. At twice the antenna's lower operating frequency, this antenna works best for short- and medium-distance contacts, but I've worked my share of DX with it, too.

Table 1
Recommended Wire Lengths and Capacitor Values*

Bands	Vertical Length (ft)	Horizontal Length (ft)	Total Length (ft)	Series Capacitor
80/40	32	64	96	≈100 pF
160/80	64	128	192	≈200 pF

*The total length is important, but the portions allocated to the vertical and horizontal members aren't critical.

The Ground System

Because the inverted L is a modified end-fed vertical, the quality of the RF ground near the feed point influences its operation. It's therefore best to use as many radials as you can, based on the available space, following the guidelines given in *The ARRL Antenna Book*.[4] My ground-radial system is modest because of limited space, and tearing up the yard certainly wouldn't win me any points with my wife. I used an 8-foot ground rod (strategically located near a sprinkler) driven into moist soil at the feed point, and three odd-length radials. One is very long (over 100 feet) and snakes along a fence, and the others are less than 20 feet long. This ground system is far from ideal, but the antenna works.

Installing the Antenna

One of the nice things about inverted Ls is

An 80-meter inverted L takes up only a quarter-wavelength of horizontal space (64 feet) and can be built to blend into the background.

---

that they fit easily on most city lots. You can run the antenna from a chimney mounted support (such as a TV mast) or tower to a tree in the yard. Or, you could run it between trees or push-up TV masts at opposite ends of the yard. The supports can be slender and lightweight because the wire's weight and wind load are small. Unlike center-fed dipoles, inverted Ls don't have heavy, unsightly coax hooked to the middle of the horizontal span. I use a pair of TV masts clamped to the chimneys located at the ends of my house to support my $^3/_8$-λ inverted L (see the title drawing).

The poles are painted sky blue and suspend insulated no. 16 stranded wire (also blue) at about 35 feet. Although the chimneys are only about 54 feet apart, the antenna works fine. I make up the difference in overall antenna length primarily by slightly angling the vertical wire away from the house. It's a good idea to space the wire's vertical section away from a building or conductive mast, even if the mast isn't grounded.

A box at the feed point firmly anchors the antenna wire at ground level, so install strain relief elsewhere in the system to keep swaying supports from breaking the wire. One popular technique often described in the literature[5] involves installing a small pulley, with a weight attached to the rope after it passes through the pulley, at the horizontal wire end opposite the feed point. As the supports sway, the weight keeps constant tension on the wire. At the top of the vertical section, the wire passes through one eye of a ceramic insulator. In my installa-

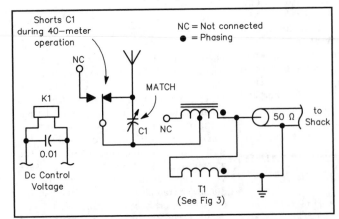

Fig 2—The resonating, impedance-matching, and band-switching circuitry required at the base of the inverted L, assuming a 50-Ω coaxial feed, no antenna tuner and a limited ground-radial system. See text for details and other feeding options. Fig 3 shows details of T1.

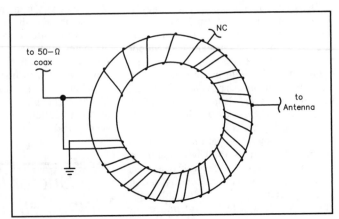

Fig 3—Winding details for constructing broadband bifilar transformer T1. You can use an Amidon FT-240-61, FT-240-43 or T-200-2 core; see note 9. The primary is 16 turns of no. 14 enameled wire, and the secondary is 10 turns of no. 14 enameled wire tapped at about the eighth turn from the feed-line end.

tion, a spring takes the place of the pulley and weight at the horizontal wire end opposite the feed point. Both techniques work well. My inverted L has survived seven winters so far, and is still going strong.

Tuning, Feeding and Adjusting the L

The inverted L can be easily matched to coax. For ease of explanation, I'll use the 80/40-meter version as an example. With a modest ground system, the antenna's feed-point impedance appears as roughly 100 Ω on both bands. If you use the dimensions given in Table 1, the antenna wire should be a tad long for resonance on 40 meters.

On 80 meters, the impedance is partially reactive, appearing as an inductor in series with about 100 Ω of resistance. Unlike 40 meters, though, the antenna's radiation resistance varies across the band (from about 70 Ω at 3.5 MHz to roughly 130 Ω at 4 MHz). To tune out the inductive reactance, you'll need to add a capacitor in series with the coax center conductor at the feed point (see Fig 2). I suggest that you use a variable capacitor to resonate the antenna at the frequency of interest. After doing this, you can replace the variable with

Remember, the transformer is there to reduce the SWR after you've resonated the antenna.

a fixed capacitor, if you like. Should you choose to stick with the variable, protect it well against weather. Before sealing the feed-point box, toss in some moisture absorbent material (desiccant).

You can make your own fixed-value, high-voltage, high-current capacitor from a piece of coax. For example, RG-8 (solid dielectric) exhibits 29.5 pF per foot and can withstand 4 kV at several amperes. Connect it by attaching one end of the inner conductor to the antenna and the braid (at the same end) to the feed system; leave the other end open. Start

You can make your own fixed-value, high-voltage, high-current capacitor from a piece of coax.

with about 4 feet of RG-8 and trim the coax for minimum SWR. (Make your cuts when you're not transmitting!) Other solid dielectric cables (such as RG-58) are also suitable for this application, *The ARRL Handbook* gives capacitance per unit length in its table of coaxial-cable parameters.

Another way to construct your own high-voltage capacitor involves sandwiching a piece of picture-frame glass between two metal plates. Richard Plasencia, WØRPV, describes how to build a variable capacitor using this technique in *The ARRL Antenna Compendium, Vol 2.*[6]

Regardless of your choice of a series capacitor, remember that it must be able to withstand at least 1 kV and fairly high currents. Use 1-A/1-kV components for 100 W output and 4-A/4-kV parts for 1.5 kW. Receiving variables work fine. Transmitting variables are hard to come by these days from commercial sources; hamfests, electronic flea markets and surplus stores are your best bet if you don't have the right part in your junk box.

Choosing a Feed Line

So far, so good. We now have a resonant two-band antenna—with a feed-point impedance near 100 Ω. What you do next depends on whether your station includes an antenna tuner. If you have a tuner, you're home free, because even if you use 50-Ω coax, the SWR on the line will be only 2:1 (higher off resonance, of course), which isn't enough to cause much additional line loss at these low frequencies. But consider the cable's breakdown voltage. Miniature RG-8 (such as Belden RG-8X and Radio Shack RG-8M) works fine at 100 W; solid dielectric RG-8 is adequate for the kilowatt level—as long as you don't try to cover the entire 80-meter band with one setting of the series capacitor (more on this later).

Feed-line SWR can be minimized by using 75-Ω coax. In fact, 75-Ω cable is the best choice. Minimizing ground-system loss by improving the radial system makes the antenna a better match for 75-Ω coax by decreasing the feed-point impedance—another good reason to install the best radial system you can.

If you don't have a tuner but your rig has tube finals, you may be able to adjust the transmitter's output network to match the feed line. Pi-network output stages (like the one in my TS-820 and most other 6146-based rigs) can handle impedances above 50 Ω pretty well. You'll likely be able to connect the 75-Ω coax directly to such a rig without an antenna tuner. Trailing-edge technology can have its advantages!

Matching the Inverted L to Coaxial Cable

If your rig won't match the coax and you don't have an antenna tuner, use a matching transformer between the feed line and antenna. For single-band operation (80 or 40 meters), you can use an electrical $^1/_4$-λ section of 75-Ω line, in series with the 50-Ω cable, to transform the ≈ 100 Ω feed-point

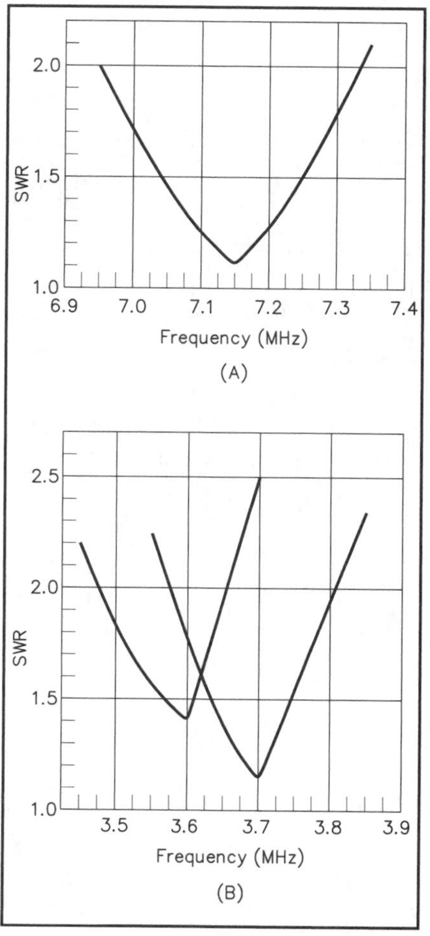

Fig 4—At A, 40-meter SWR curve using the network of Fig 2. At B, measured SWR curves for 80 meters using the same network. Curves representing two different series capacitor values are shown. A single capacitance value yields a 2:1 SWR bandwidth of approximately 175 kHz.

Fig 5—The feed-point enclosure for the dual-band inverted L at AE6C. Five high-voltage mica capacitors (at right) tune out the antenna's reactance at 80/75 meters.

impedance to 50 Ω at the end of that section. You can attach 50-Ω coax at that point or connect the matching section directly to the rig. This type of transformer is known as a *Q section*, a special case of the *series section transformer*. *The ARRL Antenna Book*[7] describes these in more detail.

For dual-band operation, you'll need to either switch transformers when changing bands or construct a 2:1 (100:50 Ω) broadband transformer (Figs 2 and 3) that operates over your frequency range of interest. The easiest way to accomplish this is to wind a broadband transformer on a toroidal core. A 2-inch powdered-iron core or 2.4-inch ferrite core can handle the legal limit if properly applied. I see no problem with ferrite-core saturation at the 100-Ω impedance level using the recommended toroids. Wind the transformer as shown in Fig 3 using no. 14 enameled wire. A kit that includes everything you need for this application is available from Amidon Associates. Although this transformer looks like a balun, it isn't. Both the input and output terminations of this transformer are unbalanced.

If you adjusted the antenna for resonance on the two bands, you need only adjust the transformer tap for minimum SWR at your rig. You should be able to cover the entire 40-meter band with an acceptable SWR, but 80 meters can only be covered in 175 kHz chunks (Fig 4). The best way to get around this is to make it easy to adjust the series capacitor, or switch in different capacitors. Fig 4B shows SWR curves representing two different capacitor values. Fig 5 shows my antenna's feed-point box, where the capacitors and shorting relay are mounted.

If you like, you can adjust the tap point on the toroid for your favorite 80-meter frequency. This will have an adverse, but minor, impact upon the minimum SWR on 40 meters.

If you're pruning the antenna and making feed-point adjustments with the aid of only an SWR meter, first adjust the total length for an SWR dip at your 40-meter frequency of interest. Then, adjust the capacitor for minimum SWR at your frequency

Make as much of the antenna as possible vertical and bend the rest over.

of interest on 80 meters. You can further minimize (or trade off) SWR by choosing the optimum transformer tap. Remember, the transformer is there to reduce the SWR after you've resonated the antenna—it won't change the antenna's match frequency. Only pruning the wire and adjusting the series capacitor can do that.

Switching Bands

To switch bands, short the series capacitor (see Fig 2). You can do this manually or via a relay. I chose a relay because I don't like going outside with a flashlight on cold, rainy winter nights—numbs the fist, you know. Relay contact spacing and size should be sufficient to handle at least the voltages and currents given for the series capacitor. (Large contacts are not necessary unless you run high power.) Ceramic relays work best for antenna switching, but they're hard to find. Hamfests, electronic flea markets and surplus stores are possible sources. At the 100-W level, you can get by using plastic-insulated relays intended for line voltage-switching applications. These are both easy to find and inexpensive.

Adding 160 Meters to the 80/40-Meter L

When I want to work 160 meters, I insert a 15-μH loading coil in series with the feed point, and short the capacitor with a relay. Initially, I used a dip meter to establish resonance. The antenna's feed-point impedance is low on 160 meters (less than 20 Ω, including ground and coil loss) so if you usually use a 2:1 transformer to step 50-Ω coax im-

pedance up to 100 Ω, bypass it. Connect the loading coil in series with the coax at the feed point. The line SWR will be high, but this shouldn't harm system performance because even a badly mismatched coaxial feed line has little loss at 160 meters. Of course, you'll need to use an antenna tuner. Alternatively, you could reverse the 2:1 transformer to step the coax impedance down closer to that of the antenna. I've never tried this, but it should work better than directly connecting the antenna to 50-Ω coax.

This is obviously a makeshift 160-meter aerial, but it works. On a good winter's night, I can work into the Midwest while running 100 W. If you've got the room, by all means put up the 80/160-meter version. It's impressive. Mitchell, KB6FPW, and I put one up in the pines last year on Field Day and cleaned out the 80-meter band with it. We didn't get a chance to exercise it on 160, though, because there was little activity. I'd love to hear from anyone who puts an 80/160-meter inverted L through its paces.

Conclusion

If you've read this far, you now realize that you no longer have an excuse to avoid the lower bands. As we slip farther down sunspot cycle 22, conditions and activity on 160, 80 and 40 meters will improve. Be there with an inverted L—and a good signal!

Notes

[1]This is because the antenna's feed-point impedance consists of radiation resistance (a constant for the antenna that represents the antenna's radiating load) and loss resistances. The higher the radiation resistance, the less toll loss resistance takes on antenna efficiency.

[2]I used an enhanced version of *MININEC* called *MN*.

[3]J. Hall, ed, *The ARRL Antenna Book*, 15th Edition (Newington: ARRL, 1987), pp 6-11 to 6-12.

[4]*The ARRL Antenna Book*, pp 3-1 to 3-6 and 3-13 to 3-14.

[5]*The ARRL Antenna Book*, p 22-1.

[6]R. Plascencia "Remotely Controlled Antenna Coupler," J. Hall ed, *The ARRL Antenna Compendium, Vol 2* (Newington: ARRL, 1989), pp 182-186.

[7]*The ARRL Antenna Book*, p 26-14.

CHAPTER EIGHT

OUR FRIEND THE TREE

By Doug Brede, W3AS   From *QST*, September 1989

The Care and Feeding of an Amateur's Favorite Antenna Support–the Tree

If your tree-supported antenna fell down, you'd care. Did you ever think about caring for the tree that holds up your antenna?

For most hams, trees are favorite antenna supports. Many radio amateurs begin their operating careers by hanging the far end of a wire up in the family's shade tree. On Field Day, resourceful hams find a hundred and one ways to get an aerial into the air; many (if not most) of these methods involve using trees as supports or aids.

During my 20 years as a radio amateur, I've used tree-supported wire antennas almost exclusively. Some of those antennas lasted several years; most didn't. Over the years, by trial and error—and because of my trade association with arborists and horticulturists—I've gained an understanding of what can (and can't) be expected of trees as antenna supports.

There are right and wrong ways to attach and maintain your tree-mounted skyhooks over the long haul. In this article, I'll share with you some pointers from two noted horticulturists who talk about attaching wires to trees. Safety is also discussed—your safety during antenna installation, and the safety of the tree.

Trees Are Alive

Few antenna supports can be classified as life forms. Trees are an exception. Tree experts usually cringe when someone brings up the idea of attaching a wire to a tree—especially when connecting a chunk of wire to its midriff (see Figs 1 and 2). The experts know that trees are made up of three basic layers: the bark, the living sapwood, and the nonliving heartwood. The bark protects the sapwood from injury. The sapwood contains the "skin and blood vessels" of the tree. If the sterile barrier between the bark and the sapwood is broken, infection can set in. Infection, if unchecked, can kill even a mighty oak within a year.

Drilling a hole through a tree causes much less trauma to the tree than wrapping something around it.

There are right and wrong ways to attach and maintain your tree-mounted skyhooks.

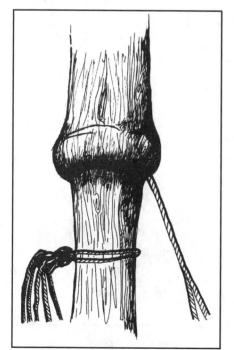

Fig 1—Attaching ropes or wires to trees can sometimes lead to major problems for the tree. Wrapping a rope around a limb or trunk and leaving it unattended will suffocate the tree and cause a distortion of growth or the death of the limb.

Fig 2—Over the years, this tree has grown around the cable of a roadside barrier. Dave Newkirk, AK7M, spotted this tree in Glastonbury, Connecticut. (*photo KC1MP*)

Some Questions and Answers about Tree Antennas

Q: A CBer in my neighborhood cut the top out of his pine tree and stuck a ground plane antenna up in it. Is this an acceptable way to mount an antenna?

A: Definitely not. Not only is this a hazardous way to mount an antenna, it essentially ends the useful life of the antenna. Topping of trees is strongly discouraged by professional arborists. Because topping removes the growing point of the tree, the tree recovers from the damage by sprouting numerous lateral buds around the top, which soon overrun the antenna.

Q: I've heard that if you fertilize a tree, your antenna will grow higher each year. True?

A: False. Although fertilizing is a desirable way to keep your tree healthy, it does not raise the height of your antenna one inch. Trees grow by extension of the apex. By the way, when you fertilize your tree, use regular garden fertilizer distributed around the drip line of the tree. The fancy tree spikes you see advertised are unnecessary because most tree feeder roots are near the surface.

Q: Is there any way to slow down the growth of a tree, so that it doesn't interfere with my antenna?

A: Some home-and-garden stores now stock growth regulators for trees. These products can be injected into the tree, dropped on the soil surrounding the tree, or sprayed on the leaves (follow label directions). Tree professionals can also perform this service. These growth regulators are used by some utility companies to reduce the need for tree trimming near power lines.

Q: Are certain types of trees better wire-antenna supports than others? What about hardwoods versus softwoods?

A: there's little difference between hard- and softwoods in their ability to hold up antennas. Conifers, because of their shape, are nearly ideal antenna supports. Avoid the use of red oaks and silver maples if possible, because they tend to rot easily if wounded. Avoid using poplars, too. In spite of their height and rapid growth, their branches are brittle and break easily.

Q: If I damage a tree during antenna installation, what should I do? Is tree replacement expensive?

A: If the damage is minor, your best bet is to do nothing. If it's a broken limb, saw the limb off cleanly, perpendicular to the axis of the branch. Never saw off a branch flush with the surface of the trunk, as this allows decay to set into the trunk. Using tree paint for repair is unnecessary (see text). In case of major tree damage, consult a trained arborist.

The answer to the second question is: Yes, tree replacement is expensive. The International Society of Arboriculture publishes a formula for calculating replacement cost of shade trees of various sizes. This pamphlet can be obtained from many tree services and libraries. Here's one point to ponder: A large, stately shade tree can add several thousand dollars in value to the property on which it sits.

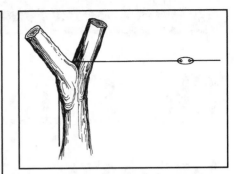

Fig 3—Most hams install tree-mounted antennas by throwing a line over a branch crotch. This should only be used as a temporary installation, because abrasion of the rope and tree results. Over time, girdling may occur leading to the loss of one or more of the branches.

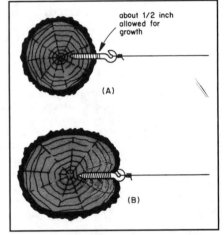

Fig 4—The best way to secure a wire to a tree is with an eyescrew mounted into the wood (A). As the tree grows and expands, however, the eyescrew will become embedded (B) and must be removed and replaced.

Trees have the same basic problems with infection as we humans do. If a tree gets a cut or gash, infection from bacteria and fungi is bound to set in. But there's one important difference between trees and humans: "Tree wounds don't heal," says noted tree expert Dr Alex L. Shigo. "People heal; when you are wounded, you have forces that fight off the infection. Trees don't have these forces to fight off infection, and every wound will become infected."

Shigo, author of the book, *Tree Biology and Tree Care*[1] notes that trees lack an immune system that fights off infection from wounds that occur from the actions of a careless climber or the attachment of an an-

tenna support eyebolt. Trees treat their wounds by walling off the infected area and isolating it from the living part of the tree. "If you cut open a tree that's 2000 years old, you'll see every injury in that tree that occurred over its lifetime," says Shigo.

Whenever you wound a tree, you weaken the tree in that spot. The walled-off wood around the wound lacks the strength of healthy wood. When attaching an antenna to a tree, it's important to traumatize the tree as little as possible. This will ensure a strong, enduring connection.

Most people believe that tree paint or shellac is the best way to treat a tree wound. "Not so," says Shigo. "Wound dressing paints just protect the microorganisms." Scientific research with tree-wound preparations have failed to show *any* benefit to the tree.

Making the Attachment

Although it's relatively easy to get a wire up into a tree, it's certainly more difficult to keep it there for the long term. Usually, annual (sometimes weekly) restring-

[1]A. Shigo, *Tree Biology and Tree Care*, (Shigo and Trees, Assoc, 2nd ed. 1989) 4 Denbow Rd, Durham, NH 08824; $52 plus shipping and handling. A companion to this book, an expanded glossary of 239 tree terms, is priced at $13. The shipping and handling charge for any single book is $3. For any combination of books ordered, the shipping and handling charge is $3 for the first book and $1 for each additional book.

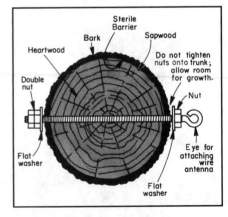

Fig 5—For heavy antenna loads, an eyebolt passed through the trunk or limb will support more weight than an eyescrew. Allow about $1/2$ inch of play between the bolt and trunk or limb. Don't tighten the bolt completely; this allows for tree growth.

> *Tree experts usually cringe when someone brings up the idea of attaching a wire to a tree–especially when connecting a chunk of wire to its midriff.*

ing is needed. It seems that trees "instinctively know" just when to drop a wire to the ground: during midwinter when the snow is high and the skip is long, or in the middle of a heated contest!

The bow-and-arrow method has become a standard of the wire-in-the-tree crew. But many other methods, slingshots, for example—even attaching a string to a golf ball and whacking it with a sand wedge—are common.

One of the easiest and most common ways to connect a wire to a tree is to throw a rope over a branch crotch (see Fig 3) and tie off the loose end. This is the main method used in temporary (such as Field Day) installations. "Doing this probably won't hurt the tree if it's done as a temporary thing," says Washington State University horticulturist Ray Maleike. But with any of these simple antenna-stringing methods, some problems for the tree (and the antenna) may develop later.

"First of all, you're not stabilizing the antenna very well with this type of setup. The other thing is that people have a tendency to forget the antenna's there. As the tree grows—as it increases in diameter—you can girdle the tree. If you've got this girdling rope or wire up there, you can actually kill that portion of the tree above the wire."

Another no-no when attaching an antenna to a tree is wrapping a wire around

> *Another no-no when attaching an antenna to a tree is wrapping a wire around the trunk. This strangles the veins in the sapwood the same way a noose around your neck would strangle you.*

the trunk. This strangles the veins in the sapwood the same way a noose around your neck would strangle you. "It's important not to wrap anything around the trunk," says Maleike.

Many commercial nurserymen wrap stabilizing ropes around newly transplanted saplings to keep them from falling over. Recently, however, this practice has been questioned because of the restrictions these ropes place on the growth of the tree. People forget about these ropes; some remain on trees for years after transplanting.

Encasing the stabilizing (or antenna) wire in rubber or plastic hose is not the answer either. "Wire wrapped in hose is just as injurious to the tree as the bare wire itself," says Shigo. "If you remember your basic physics, you're applying the same number of pounds of force to the tree with or without the hose." Shigo recommends that if you must wrap something around the trunk of a tree, use a wide fabric strap to do the job.

Two methods have emerged among leading horticulturists as the preferred way to attach a wire to a tree. For light antenna loads (eg, the end of a dipole), a threaded eyescrew (Fig 4) is the method of choice. Simply drill a hole into the tree about $^1/_{16}$ inch smaller than the screw diameter, then twist in the eyescrew. Be sure to select a cadmium-plated eyescrew threaded for use in wood. A thread length of 2 or 3 inches should secure most antennas. Allow about $^1/_2$ inch of space between the trunk and the eye; this allows for outward growth of the tree with time.

For stouter antennas, such as multielement wire beams, another method for securing wires to trees is recommended. This procedure involves using an eyebolt longer than the tree diameter, drilling clear through the tree and securing the eyebolt

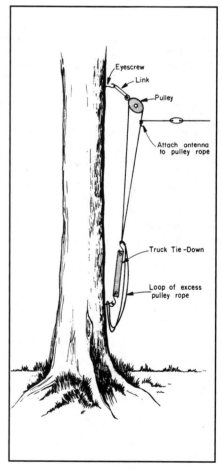

Fig 6—By using a pulley, raising and lowering the antenna for repairs can be done without the need to climb the tree. Flexible truck tie-downs can be used to apply tension to the antenna. (Early editions of *The ARRL Antenna Book* show a weight used to provide the required tension. A weight swinging from a tree can be hazardous.) Loop the excess pulley rope to a second eyescrew, in case the tie-down fails.

Fig 7—A profesional arborist uses a safety belt and rope when climbing trees. Hams should take similar safety precautions. (*A. Douglas Brede, W3AS, photo*)

on either side of the tree with round washers and nuts (see Fig 5).

Drilling a hole through a tree causes much less trauma to the tree than wrapping something around it. Much of the core of a tree is dead tissue, used mainly for physical support. Although there will be some wounding of the tree at the site of the bolt or screw, such wounding will be far less than that which occurs from wrapping a wire around the trunk.

Over time, either type of eyescrew connection will have to be replaced. "If these fasteners are left on the tree for a long time, the fastener will eventually become embedded in the tree," says Maleike. "You're go-ing to have to pull these fasteners out and replace them every now and then." Maleike recommends replacement of tree eyescrews every 5 to 8 years as the tree matures. Commercial arborists use drive fasteners for securing wires to trees; drive fasteners are similar to eyescrews. "These fasteners keep the wire away from the tree, allowing the tree to grow out to it," says Maleike. Drive fasteners are used for securing lightning rods and their accompanying wires to trees. The use of drive fasteners is common in the Midwest, where lightning strikes to trees are common. You may have to shop around to find drive fasteners—try calling tree-care services in your area.

It's easier to periodically service a tree supported antenna if a pulley is used (see Fig 6). Raising and lowering the antenna for repairs can be done without the need to climb the tree each time. I use a flexible truck tie-down to provide tension to the antenna.

Your Safety in Trees

A fall from a 40-foot tree is just as dangerous as a fall from a 40 foot tower. Yet, many times you see hams scaling trees with no safety equipment! Wear a tower-climbing safety belt for all tree climbs (see Fig 7). Commercial arborists take the matter of safety one step further: They lob a rope over a tree crotch just above the height at which they'll be working. Then they tie the rope to their safety belt. The loose end of the rope can be held by a helper on the ground.

Be sure to use a good quality rope that is heavy enough to support your weight. Before use, inspect the rope for wear. Arborists prefer to use hemp rope rather than nylon, because hemp rope stretches less.

When you're climbing a tree to attach a wire, always have a buddy on the ground available to fetch tools or summon help in an emergency. Be sure your buddy wears a hard hat; tools or branches dropped from even a moderate height can be dangerous.

Trees lack an immune system that fights off infection from wounds that occur from the actions of a careless climber or the attachment of an antenna support eyebolt.

As an alternative to doing it yourself, consider procuring the assistance of a professional to install your tree antenna. A professional can clear away interfering branches and secure an eyescrew in short order. Professional tree trimmers generally work in pairs. They use a ladder or bucket truck to get up into the tree, and then they free-climb throughout the tree. A safety rope, saddle, and safety belt are worn. "A figure that I heard about how much this runs is about $50 an hour," says Maleike. Most antenna tasks can be done by professionals in about an hour.

Summary

Keeping your station in good operating condition is—or should be—a fundamental practice of every radio amateur. Part of that practice includes annual inspection of your antenna system. If trees are a part of your antenna system, take a good look at them. Are you keeping them healthy?

By Wade A. Calvert, WA9EZY　　　　　　　　　　　From *QST*, June 1991

The EZY Launcher

Lazily lift a line to a lofty limb with this little launcher.

Would you like to raise your 75-meter flattop? Do you get a queasy feeling whenever you consider how you're going to get that nylon rope over that "perfect" limb at 50 feet? Try the EZY Launcher!

The EZY Launcher is made from a Zebco model 202 spin-casting reel (complete with fishing line), a Marksman slingshot, a piece of flat steel stock, a wood dowel, two small hose clamps and some no. 10 hardware. I obtained all the parts at a local hardware store for less than $15. If your hardware store isn't as well stocked, chances are you can get the fishing reel and slingshot from a sporting goods, discount or catalog-sales store. The accompanying photographs show how easy it is to build an EZY Launcher.

Construction

A view of the disassembled EZY Launcher is shown in Fig 1. Table 1 contains the parts list. The U-shaped bracket is formed from a piece of $^3/_4$-inch-wide, $^1/_8$-inch-thick steel (or aluminum). Each leg is approximately $2^3/_4$ inches long, with a $1^3/_4$ inch gap between the legs. Adjust this gap to comfortably fit your hand and allow your thumb to operate the fishing reel's release button at the rear. To determine the gap width, grasp the slingshot in your shooting hand. Measure the distance from the side of the slingshot handle nearest you to the second knuckle of the middle finger of your shooting hand; to this dimension, add about $^1/_8$ to $^1/_4$ inch. Cut the metal stock long enough to allow for the leg lengths, the bracket's horizontal portion and the bends at each leg. Using a vise, bend the flat stock at the required gap distance to make the horizontal part of the bracket.

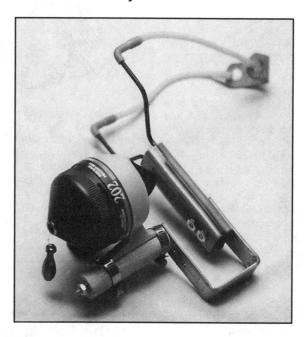

What do you mean, storm the castle. . . I thought we were going to launch my dipole!

Then, cut the legs to length.

Drill two $^3/_{16}$-inch-diameter holes at the top of one of the bracket legs on the centerline. Locate the first hole $^3/_8$ inch from the top of the leg. Make the second hole $^1/_2$ inch down from the first hole. Deburr the holes and countersink them on the outside of the bracket. Drill a single $^3/_{16}$-inch diameter

hole $^3/_8$ inch down from the top of the other leg on its centerline. Deburr the hole and remove any sharp edges from the bracket.

Place the slingshot against the bracket and check for grip clearance. Drill two mounting holes in the slingshot handle. Position the first hole $^3/_8$ inch up from the bottom of the wooden part of the handle

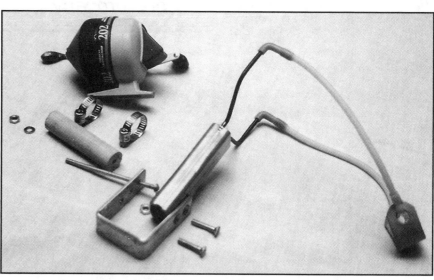

Fig 1–A view of the disassembled EZY Launcher.

Table 1
Parts List

Qty	Item
1	Zebco model 202 spin-casting reel (or equivalent)
1	Marksman Co model 3010 slingshot (or equivalent)
1	$1/8 \times 3/4 \times 12$-inch piece of flat steel (or aluminum) stock
1	$3/4$- $\times$ 3-inch hardwood dowel
2	$3/4$-inch hose clamps
2	No. 10 $\times$ 1-inch flat-head machine screws
1	No. 10 $\times 3^{1}/2$-inch round-head machine screw
3	No. 10 nuts
1	No. 10 flat washer
1	No. 10 lock nut
1	$1/2$-oz sinker

on its centerline. Locate the second hole $1/2$ inch from the first. These holes should line up exactly with the holes in the bracket. Attach the slingshot to the inside of the bracket using two no. 10 $\times$ 1-inch-long flat-head machine screws. To keep the screw ends flush with the nuts, I used no washers beneath the nuts.

Drill a $3/16$-inch-diameter hole lengthwise through the center of a 3-inch-long dowel. A drill press comes in handy here, but a satisfactory job can be done with an electric hand drill and reasonable care. Use a $3^{1}/2$-inch-long no. 10 machine screw to attach the dowel to the bracket leg opposite the slingshot. The bolt head should be inside the bracket. Secure the dowel using a no. 10 flat washer and nut.

Attach the fishing reel to the dowel using two small hose clamps. Grip the slingshot and rotate the dowel until your thumb can rest comfortably on the release button. Slide the reel forward or backward if necessary.

Operation

Safety First!

Remember: A slingshot is not a toy! Wear eye protection and take any other precautions necessary to ensure the safety of people and property. Never shoot near power lines. Practice shooting in a flat, open area until you're confident in your ability to shoot predictably.

Remove the metal ring at the end of the fishing line. Make a 1-inch loop of line at the end. Carefully deburr the eye of a $1/2$ oz sinker. Push the line loop through the sinker eye, pass the loop around the sinker body and pull it tight. This method of fastening the sinker to the line makes it easier to get the sinker off the line and provides a means of connecting light nylon line once

> *Remember: A slingshot is not a toy! Wear eye protection and take any other precautions necessary to ensure the safety of people and property.*

the sinker has passed over the tree limb and returned to ground. (Please have regard for the trees. Doug [W3AS] Brede's article[1] is recommended reading.—*Ed.*)

When you're ready to shoot, wind the sinker tightly against the reel. Make sure that the reel crank is down and out of the way of the shot. If it isn't, press the release button while holding the sinker in place with your free hand, and wind the crank until the sinker is tight against the reel and the crank is down. Position the sinker in the pouch with the sinker eye pointing down. Take short shots until you get the feel of the Launcher. Control the shot distance with the release button. The button not only releases the line for casting, it also acts as a brake when depressed further. You can set the degree of drag by adjusting the black adjustment wheel on top of the reel. When rewinding the line, apply a small amount of drag to it using your thumb and forefinger.

That's it! Have fun with the EZY Launcher. It's sure to make your antenna raising easy.

[1]D. Brede, "The Care and Feeding of an Amateur's Favorite Antenna Support—the Tree," *QST*, Sep 1989, pp 26-28 and 40. [It's the first article in this chapter.]

A New Way to Tree A Wire

Trees are beautiful in many ways. They provide shade and color; they act as air purifiers and air conditioners. They also do a great job of supporting wire antennas.

Rarely are trees used to their full advantage as antenna supports, because it's hard to get wires very high in them. There is a limit to how high you can safely climb a tree. It isn't cost effective for many hams to rent "cherry pickers" to reach up to the tops of most trees. The bow and arrow, with the arrow's point blunted and weighted, is one good way to get wires over treetops. It's also an excellent way to get into trouble with neighbors (broken windows, skewered cats, and such).

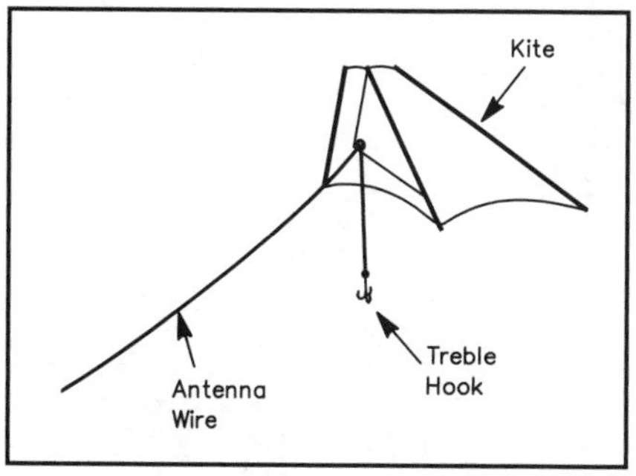

The Kite

Every March and April we see demonstrations of an elegant, simple, cheap and safe device that's well suited to getting wires just where we want them. Kites, like trees, are pretty to look at. And the two just naturally seem to go together. Kids know that. Kites can be found in trees all year round; sometimes they stay in the branches for months or even years. In anger, the little child, after losing a kite to a tree (again), might exclaim, "May as well just put the thing up there on purpose!" Exactly my idea.

Even a dime-store delta kite flies quite stably in a light breeze, and believe me, if you want to tree it, both the kite and the tree will eagerly accommodate you. All you need is one of those little three-hook things for securely catching fish [properly called a treble hook—*Ed*], available at any tackle shop, placed on a short string dangling from the kite tether point (Fig 1). You can be sure that once your kite is treed, it'll stay treed for as long as the line and the hook and the tree will last, barring major windstorms.

This method doesn't allow for error in measurement of the wire. It also is not very good for stringing heavier gauges of wire, but for low-power and moderate-power amateur transmitters, heavy wire is not necessary anyhow. I have found that no. 18 or no. 20 stranded copper wire is entirely adequate for most installations up to about 500 watts.

The treeing-of-the-kite-on-purpose (TOTKOP) method works best for "ran-

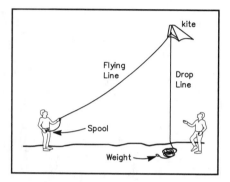

Fig 1—Simple method of freeing a kite for temporary long-wire or random-wire installations. When the kite hits the tree, it'll stay, almost every time.

dom" wires that you simply run out the window to a nearby tree. It is especially convenient for apartment dwellers. No one needs to know that the treed kite has anything to do with your wire antenna. The "stealth factor" can be enhanced by treeing the kite late at night.

Safe Kite Flying

Use nonconductive line whenever possible, but no matter what type of line you use: Don't fly it where the line can possibly contact power lines. *You can be killed!* Also, don't fly a kite in a thundershower, or even when one is forecast or threatening. Use care when flying a kite over parking lots or buildings: You may tangle up the whole neighborhood! Your antenna must never pass over or under utility wires, and it should not be placed where power or phone lines can contact it if a windstorm breaks your antenna wire or a power or phone line. In short, use the same precautions with this method of antenna erection as with any other method.

For More Permanent Antennas

The TOTKOP method is good for mountain topping, Field Day, and other temporary purposes. For more secure installations, I have developed a safe and elegant way of getting lines over trees and placing the line just where I want it, while allowing adjustment of antenna length and slack.

The initial setup is shown in Fig 2. The best type of kite is a fairly large, stable de-

Fig 2—Initial setup for permanent method of putting a line over a tree.

vice, such as the delta or parafoil. The delta flies at a high angle (about 70 degrees with respect to the horizon) in winds of about 10-20 miles per hour. The parafoil flies at a lower angle (about 45 degrees) in winds of 15-30 miles per hour.

This method does not usually involve loss of the kite to the tree, although trees have a voracious appetite for kites. (You probably learned that as a kid!) It pays to have a spare kite handy, just in case. Conditions are ideal when the wind is steady at about 10-15 miles per hour and you can use a delta with a wingspan of 6 feet or so. The flying line should have a breaking strength sufficient for the antenna you wish to support. The drop line can be a lighter, cheaper string, weighted with a large nut or bolt, and long enough so that you can reach it however high the kite is flown. A good rule is to make the drop line twice as long as the tree is high.

For easiest placing of the line, two people are needed. One flies the kite and the other manages the drop line. The kite can be let out and flown high enough so that the flying line clears the tree you want to run the antenna support over. The person flying the kite (say it's you) should move into position upwind from the tree, while the drop-line person moves around behind the tree (Fig 3A). At this time the kite will be considerably higher than the top of the tree, and the flying line will just clear the treetop.

The drop-line person starts pulling on his line to bring the kite down (Fig 3B) without

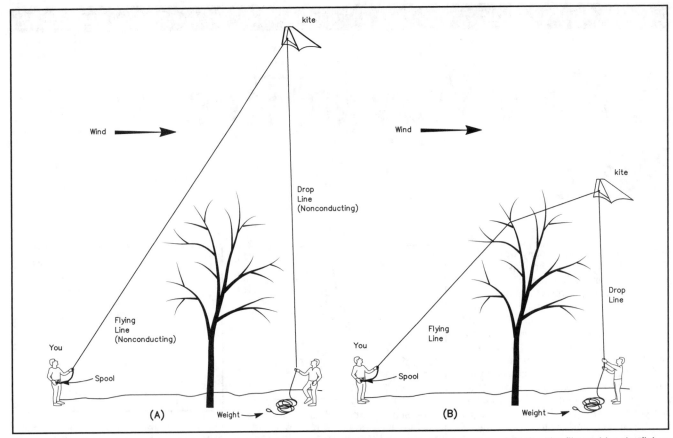

Fig 3—At A, the line is in position for treeing. At B, one person pulls the kite down via the drop line, while the kite flier guides the flying line exactly where wanted.

letting it get caught in the tree. If necessary, you should let out a little more line to ensure that the kite stays away from the tree, since kites and trees display an attraction that varies inversely according to the square of the distance between them (Murphy's Law). Be certain the kite does not get within the radius where the tree claims it forever.

It isn't hard to get the flying line right into the niche you want. Place it as high as possible in the tree, but choose a branch strong enough to support the tension on the wire, once the antenna has been strung. If the branch isn't strong enough, it will break, and the line will settle lower in the tree. Keep this in mind as you select your niche, so that if the branch breaks, the antenna will still be up fairly high and you won't have to restring the line.

Securing The Line

Have your partner pull the kite all the way down as you let out flying line from the spool. The kite can be retrieved (no small matter if you have spent upwards of $25 for it) and the line will be in a good position for you to attach the antenna wire and pull it up.

You or your partner should attach the line to the tree by tying the line around a protrusion such as a sawed-off branch, or around the tree trunk. You might want to secure the line up 10 or 12 feet from the ground, using a stepladder, so no one can easily get the thrill of untying the line and watching your antenna meet its demise.

Attach an insulator to the other end of the line, knotting the line securely the way you were taught in the Scouts. This means you will have to cut the flying line. With luck and planning, you've bought 500 or 1000 feet of line, so you can string lines over two or more trees as needed, without running out. (You didn't? Murphy strikes again!)

Then attach the antenna wire to the insulator. You can pull the antenna up from the other side of the tree, cutting off the extra line. The result will, when the whole antenna is finally erected, look like Fig 4. You can put a strong door-closing spring in the line just past the insulator, to allow for tree movement in the wind. Since you have attached the line via a rather high point in the tree, this movement may be considerable. The wire will break eventually, of course, no matter what precautions you take. Wire antennas usually don't last long, so work with this fact in mind.

A reminder: Use nonconducting line whenever possible. Ham radio is fun, but it's not worth your life. It makes sense to minimize the chances of electrical disaster by every possible means. Another closing bit of advice: Keep an extra kite handy.

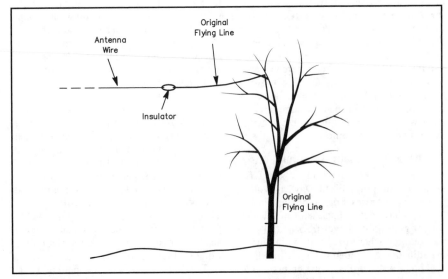

Fig 4—The final antenna installation (one end of the antenna).

By Chuck Hutchinson, K8CH

From *QST*, September 1984

A Tree-Mounted 30-Meter Ground-Plane Antenna

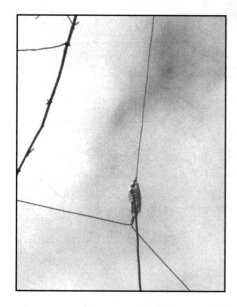

Work DX with this inconspicuous low-cost and easy-to-build wire antenna. It's great also for portable operation.

A tree-mounted vertically polarized antenna? Sound silly? Well, it's not! Perhaps textbooks do not recommend it, but I'm having good luck with a 30-meter "ground plane." The antenna did not cost much, is inconspicuous and works quite well for DX QSOs.

The evolution of this idea began more than a decade ago. My friend Al Francisco, K7NHV, was a doctoral candidate at Michigan State University. Al was living on campus and wanted to get on the air from there. University rules did not permit outside antennas. How would he solve this dilemma?

Al's solution was simple, and it worked! He ran a piece of RG-58/U cable from a bedroom window to the ground. From there, he slit a shallow trench to a nearby tree and buried the cable. At the base of the tree a couple of radials were soldered to the coax-line braid, then buried. Another piece of wire formed the main radiator of his 20-meter vertical. You had to walk right up to the tree to see it! Even more amazing to Al and his friends was that it worked about as well as you would expect for any 20-meter

vertical with only a couple of radials. It was a good idea, the kind one doesn't forget!

Several years later I moved to a new location. In short order, a modest tower with a triband beam sprouted from the back yard. Dipoles for the 40 and 80-meter bands were hung in the trees. It was possible to work DX on 40 meters, but the low dipole worked too well on short skip; a vertical antenna would be a better alternative. I knew from experience that a vertical with 16 or more radials would be a good performer.

Where to put the antenna? That's when Al's idea came back to mind: Use a tree to support the vertical radiator! This time a TV mast, not wire, would serve as the vertical radiator. About 30 feet from the corner of the house stood a full-grown walnut tree. A look at the tree confirmed that three branches were perfectly situated for supporting a 40-meter vertical.

Later that day, with the help of my teen-age son, the vertical was eased into position close to the tree trunk. A length of treated 4- × 4-inch lumber was buried to serve as the base insulator.[1] A large nail held the vertical mast in place. My son (Bryant is a good tree climber) wrapped a short length of clothesline around the mast at each branch. Next, he tied four square knots. The loose ends were then wrapped around and tied to the branches. We fimished the job by burying 32 radials around the mast and running RG-8/U cable into the ham shack. The results were great! For the next three years the antenna served me well. Many choice DX tidbits were snagged while using that vertical antenna.

An Antenna for 30-Meter DXing

On the day the 30-meter amateur band

was opened by the FCC, I used a Transmatch and 40-meter dipole to make a few contacts. Later, I tried an 80-meter dipole. Both were okay, but each worked too well on short skip. Stations within a couple of hundred miles were very loud— not the best situation for DXing!

A vertical antenna would cure that. . . perhaps another version of the K7NHV special. There was a serious problem, however. Limited space and rocky soil meant that a good radial system would be almost impossible to realize. Rats! But wait, why not build a ground-plane antenna? I could make it all from wire, and two radials should be sufficient. I went to work.

I reached for my calculator and came up with the proper length using the formula:

$$\ell \text{ (feet)} = \frac{234}{f_{MHz}}$$

[1]m = ft × 0.30483; mm = in × 25.4
kg = lb × 0.4536

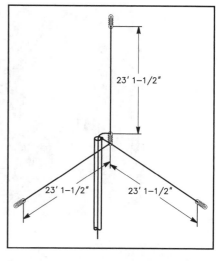

Fig 1—Dimensions and construction of the 30-meter ground-plane antenna.

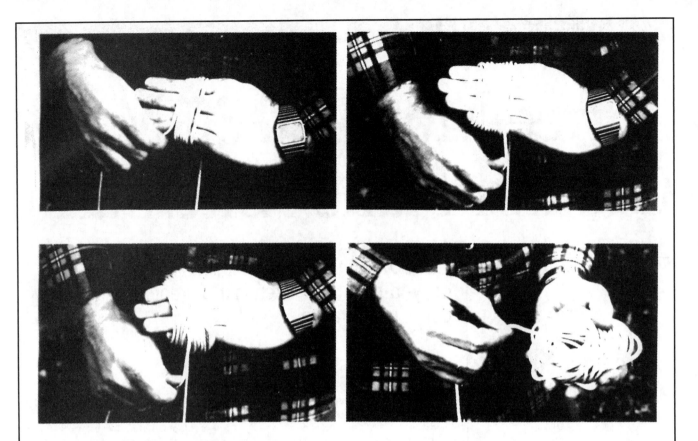

Fig 2—Chuck, K8CH, demonstrates how Ben Hassell, W8VPC, prepares a line for throwing. When the weight is thrown, the line should come freely out of the middle of the ball.

The vertical portion and each radial of the ground plane should be 23 feet 1½ inches for resonance at 10.12 MHz. The wire and insulators were on hand. It did not take long to assemble the antenna as shown in Fig. 1.

This time, the perfect tree was found at the back of the lot. Beneath it grows a lot of brush—it makes it hard to work around, but great for camouflage. Only one obstacle stood in the way of speedy installation. How to get a line through the crotch 40 feet above the ground?

Getting a Line into a Tree

Many methods are used to get an antenna support line into the "right" crotch of a tree. You could use a bow and arrow or a slingshot. A strong person can throw a no. 18 nylon line, with a proper weight on the end, to about 40 feet. That's good enough for this project.

There are a few tricks you should avoid learning "the hard way." First, make sure there is nothing breakable within throwing range. Unless you are extremely fortunate, at least one throw will go astray. Second, secure the free end of the line so it does not end up out of reach in the air. This is particularly frustrating when you have just managed to "hit the target."

I like to use an 8-inch adjustable wrench for a throwing weight. It has just about the right heft, and it is nice and smooth. If you

Adapting the Tree-Mounted Ground Plane for Other Bands

You say you like the idea in the article, but you want to use it on another band? No problem!

In the decade and a half since this article first appeared in print, many hams have used this idea for an easy-to-install antenna. For the most part, the antennas have been used for temporary or portable operations—such as Field Day. That is as it should be. If you've read the first article in this chapter, you understand that a line run through a tree crotch works well as a temporary support, but should not be used in a permanent installation.

To use this antenna on other bands, start with the lengths given in the table. You can make the radials a bit longer and adjust their lengths for best match at your favorite operating frequency. Yes, you can tune the ground plane by adjusting only the radials. It works as well as adjusting the vertical radiator, and it's much easier to do.

—*Chuck Hutchinson, K8CH*

Leg Lengths for Dipole and Ground-Plane Antennas

Operating Freq (MHz)	Length feet	inches
3.50	66	10
3.75	62	5
4.00	58	6
7.00	33	5
7.20	32	6
14.00	16	9
14.25	16	5
18.11	12	11
21.00	11	2
21.30	10	12
24.93	9	5
28.00	8	4
28.40	8	3

don't like to gamble, use something else. For this project, I found a floor flange that weighed a bit less than a pound. It worked quite well.

When you miss your target, as I frequently do, don't try to pull the throwing weight back over a branch. Take it from one who knows the indignity of viewing a beautiful, shiny adjustable wrench swinging in the breeze from a rotten old tree branch. It takes only a few seconds to let the weight fall to the ground and then pull the line on through the tree. That way you get another try with the same weight. Trees look ugly when they are decorated with dangling wrenches, transformers, pipe fittings and rocks!

Make sure your throwing line does not tangle, or you will have a mess. Ben Hassell, W8VPC, uses a method that is particularly effective in brush or tall grass. Ben lays the throwing weight on the ground. He then scramble-winds the line into a ball around his fingers (Fig. 2). The end of the line tied to the weight comes out of the middle. For a right hander, lightly grasp the ball in the left hand. ("South Paws" reverse hands.) With the right hand, lift the weight by the line and swing it 'round and 'round. Let the weight fly in the direction of the "right" crotch. Success may require several attempts, but keep trying.

Working on a mowed lawn is much easier. Lay the line out in front of you in large S shapes. Make sure there are no twigs or stones for the line to catch on. You want the line to feed smoothly from the ground as the weight goes flying on its way.

Final Steps

It took a few tries to get the line through that crotch at 40 feet. After that, things went easily. I tied the top insulator to the end of the line and hauled it up to just below the

Fig 3—One method for securing the down-haul line to a tree. Excess line is neatly stored out of the way.

crotch. A doubled-up section of the down-haul line was wrapped around the tree at head level, and tied. Excess line was secured as shown in Fig. 3. Shorter lengths of line were tied to the radial ends and were secured to convenient trees above head level.

After several weeks of operation, I find myself well satisfied with the tree-mounted ground-plane antenna. It works at least as well for DX as does a dipole at 60 feet. For stations less than a couple of hundred miles away there is pretty good rejection. (The dipole works for them.) There is no "dent" in my pocketbook, but perhaps best of all, the antenna is almost invisible. Why don't you give it a try?

Trees and Verticals

Many hams use trees to support different antennas. However, very little data has been published about the effect of the trees on antenna performance. Moxon[1] mentions "It is known that tree trunks can have a serious effect on vertical elements up to distances of at least a few feet." Devoldere writes:[2] "...trees can be very lossy elements in the near field of a radiator."

In other antenna references, including *The ARRL Antenna Book*, we haven't found any more than a general warning that "trees may introduce attenuation." The consensus appears to be, however, that the effect on vertical antennas is more severe than on horizontal antennas. In pursuing problems with two vertical antennas hanging in trees, we've made some observations that may shed a little more light on the subject. (An HP 1606A RF impedance bridge was used for all measurements.)

The first case involves two 60-foot vertical cages, top loaded, hanging in separate oak trees and fed as an array. At resonance (1.83 MHz), the first element exhibited a resistive component of 37 Ω, while the second, identical, with virtually the same ground system and only 100 feet away, measured 75 Ω. Each element, modeled with *ELNEC*, should have a resistive component of about 25 Ω.

The second case involves a single quarter-wave monopole hanging in a tall pine with seven radials raised 15 feet off the ground. The antenna was spaced 12 inches from the trunk of the tree at the bottom, and about three feet at the top. We measured the impedance of this antenna and found a resistive component at resonance (3.74 MHz) of 50 Ω. The Christman model of this antenna reported a resistive component of about 32 Ω, so there were 16 Ω of unexplained resistance. Our first experiments were made on this antenna in July 1991.

Our first experiment involved moving the base (current maximum) of the antenna away from the tree by about 15 feet. This made no difference to the resistive component at resonance. Sometime later, we got around to moving the top of the antenna away from the tree. We found that by moving the top only 6 feet away from the tree (the bottom remained 12 inches from the tree) the resistive component decreased from 50 to 35 Ω, and the resonant frequency increased from 3.74 to 3.77 MHz. Apparently, coupling to the tree at the voltage maximum was reduced to the point that the large resistive component was effectively eliminated.

We now had something to look for on the array, so we compared the arrangement of the elements at the top and found, as we now expected, that the first element was near only one large tree and cleared the large branches and trunk by more than six feet, while the second element was not only surrounded by large oak trees, but the top was very close to the trunk, and the insulated wire that provided the capacitive hat was tight against a large limb. When we moved the top about six feet away from the trunk, the resistance dropped from 75 to 47 Ω. The capacitive wire was still only two feet from the trunk, so we added another rope and moved it about six feet from the trunk. The resistance then dropped to 45 Ω. That was the lowest we could get with this element, probably because it was surrounded by trees; the top is about 10 or 15 feet from three large trees. By adding two more raised radials, the resistive component was finally reduced to about 42 Ω, still rela-

tively poor, but it will have to do for now.

The implications are that trees can introduce significant losses to a vertical antenna, particularly when the top (or voltage maximum) is close to large trunks or limbs. This problem may even extend to horizontal antennas where an end is close to a large tree or to other antennas that have voltage maxima very close to large trees. Small limbs (less than two or three inches in diameter) did not seem to add loss. Because hams make frequent use of trees as antenna supports, we think this would be a good subject for further study, either by experiment or by modeling. We expect, however, that only the full *NEC* (numerical electromagnetics code) program could model the dielectric nature of the trees.
—*Clay Whiffen, KF4IX, and Ben Zieg, K4OQK*

> *Trees can introduce significant losses to a vertical antenna, particularly when the top (or voltage maximum) is close to large trunks or limbs.*

> *Small limbs (less than two or three inches in diameter) did not seem to add loss.*

[1]L Moxon, "HF Antennas for All Locations," (Potters Bar: RSGB, 1982), p 194.
[2]J. Devoldere, *Low-Band DXing*, (Newington: ARRL 1987), p 2-40.
[3]Christman, "Elevated Vertical Antenna Systems," *QST*, Aug 1988, pp 35-42. Also see Feedback, *QST*, Oct 1988, p 44.

By Mark Weaver, WB3BJF   From *QST*, November 1995

A Four-Band "Tree" Vertical

If the thought of a high-visibility HF antenna leaves you cold, it's time to branch out and get to the root of the problem...

I live in a townhouse on a small lot in a neighborhood where no outdoor antennas are allowed. That's a fairly typical situation these days. So if I want to operate on the HF bands, am I resigned to an attic dipole or some other indoor compromise? No way! Believe it or not, I'm the proud owner of a four-band full-sized vertical antenna, and it's sitting right in my front yard. And the best part of all is the fact that my antenna is virtually invisible. No Klingon/Romulan cloaking devices here; just old-fashioned ingenuity.

The Concept

I tried an attic dipole and had nothing but problems. RF got into everything! It got into the TV and the kids howled. It got into the telephone and my wife howled. The antenna also picked up every kind of noise from my computer, TNC and any other electronic devices in the house.

If you can see a tree anywhere on your lot, you've just found a home for your next antenna–and it will probably outperform any indoor design.

One day while staring out my front window, dreaming of 100-foot towers and stacked Yagis, my gaze fixed upon a solitary 20-foot tree in my front yard. Wait a minute! I can run a 15-foot hunk of wire up the side of that tree! That's almost a quarter wavelength on 20 meters! But what about 40 meters, one of my favorite bands? I decided to worry about that later. Thus was born my four-band "tree" vertical.

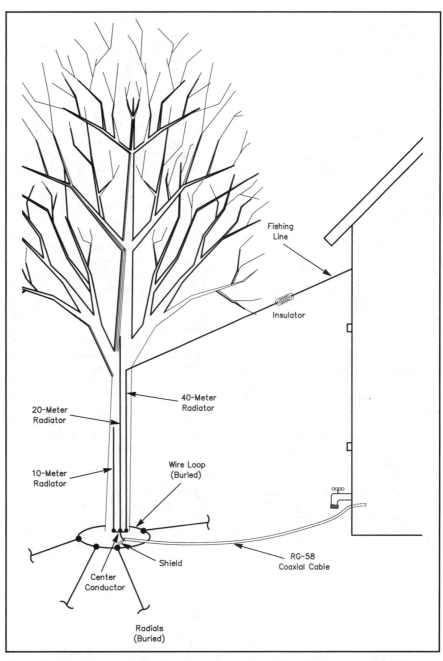

Fishing Line

Insulator

40-Meter Radiator

20-Meter Radiator

10-Meter Radiator

Wire Loop (Buried)

RG-58 Coaxial Cable

Shield

Center Conductor

Radials (Buried)

Fig 1—Run the three antenna wires along the trunk and then, if necessary, bend them along the branches. My 40-meter wire is so long that it leaves the tree altogether and attaches to my window frame. Use at least six radials for your ground system, more if you have the space and the patience to place them beneath the soil.

> *Is this antenna "optimal?" Far from it. No doubt there is some RF absorption by the tree. The point, however, is that this antenna solved my problems.*

> *The best part of all is the fact that my antenna is virtually invisible. No Klingon/Romulan cloaking devices here; just old-fashioned ingenuity.*

My idea isn't new, although the application may be unique. The antenna is comprised of three quarter-wavelength wires (for 10, 20 and 40 meter), snaking up the side of the tree, more-or-less in parallel, all soldered together at the bottom to the center conductor of the coax (see Figure 1). Several radials are then soldered to the ground braid of the coax. But how do you get a quarter wavelength wire for 40 meters into a 20-foot tree? That's over 30 feet of wire! Easy. Bend the wire at the halfway point and run it to an upstairs window of your house, or some other convenient support. When you're finished you'll have an inverted L on 40 meters, a vertical with the top bent over so it looks like an upside down L. The 40-meter inverted L also works on 15 meters, where it's a ³/₄ wavelength.

Construction

Cut three pieces of wire at quarter wavelengths on 10, 20 and 40 meters using the formula:

Length (feet) = 234 / Frequency (MHz)

Choose frequencies that are in the middle of the bands, or your favorite band segments. I recommend #26 enameled wire available at Radio Shack. It's strong and difficult to see.

Examine your chosen tree and the surrounding area. Make absolutely certain that you're not near any power lines. If you see power lines running through the branches, find another tree.

Once you've selected your tree, run the wires up the side of the trunk. If you're an experienced tree climber, work your way up the branches taking the wire along as you go. The alternative is to use a ladder, but make sure you have someone on the ground hold it for you. If inquiring minds want to know what you're doing, explain that you're trimming the tree and/or inspecting the leaves. You can attach the wires to the tree with loops of fishing line, or any other low-visibility means.

The idea is to get the wires as high as possible. You'll probably have to bend the 20 and 40-meter antenna wires, depending on the height of your tree (and your desire to climb it!). The tree in my yard is about 20 feet from the house, so the end of my 40-meter antenna reaches to an upstairs bedroom window. If you use your house as a support, you need to insulate the antenna wire so it won't come in contact with metal siding, storm windows or whatever. I loosened a screw in my metal window frame, tightened it down on a piece of insulated wire and tied the antenna to the wire.

Now build your ground system. Take bare copper wire, preferably something thick like #14, and loop it around the bottom of the tree at ground level. Solder several radial wires to this loop and run them out into the yard. I only used six radials at various lengths, making each one as long as possible. Bury the radials about an inch beneath the soil. (Do this at night if you live in an antenna-restricted area. If anyone asks, just tell them you're checking the lawn for grubs.) The radial wires don't have to travel in straight lines. Zigzag them as much as necessary to fit the available space.

Now install the transmission line. My townhouse, like many, has a water spigot on the front. That means there is a hole through the house for the water pipe to pass through. There was enough extra room in this opening to pass a length of RG-58 coaxial cable. You may need to bury this coax between the tree and the house, so make sure to buy cable that's made specifically for burial in soil.

Back out at the tree, solder all three antennas to the center conductor of the coax and solder the coax braid to the copper radial loop.Weatherproof the coax connections. I used Radio Shack "Outdoor RF Connector Sealant" (part no. 2781645). Cover the copper radial loop with mulch or soil so it won't be visible. I planted pansies around mine and it looks very nice.

Testing

An SWR meter is all you need for testing. If you measured the antenna lengths correctly, the SWR will probably be no higher than 2:1. If you want it lower, add a few inches of wire to the antenna, or trim it as the case may be. If you have an antenna tuner, you don't have to bother with tuning unless the SWR is grossly out of whack. Simply adjust the tuner for a flat 1:1 SWR. Because your transmission line is likely to be short, an elevated SWR isn't as bad as it seems. On 15 meters you're using the 40-meter antenna on the third harmonic. This means that your SWR might be high, but the tuner should be able to take care of it.

Results

Is this antenna "optimal?" Far from it. No doubt there is some RF absorption by the tree, and the radiation patterns probably look like abstract art. I'm sure that some RF is being used to heat the coax when the SWR is high.

The point, however, is that this antenna solved my problems. It works well and is far enough from the house that I no longer have complaints about TVI and telephone interference. Signals from my computer and TNC are but distant memories.

Many operators are astonished when I describe my antenna. They can't believe that my signal is so strong. When conditions are decent, I even work a fair amount of DX. I've also managed to use the system on 30 and 17 meters with good success.

As far as visibility is concerned, you can't spot the antenna unless you walk right up to the tree. Even then, you need to know what you're looking for. So far it's been completely disregarded by the spies from the home-owner's association.

Take it from me: If you live in an apartment, townhouse or condo, you can get on the HF bands with a full-sized antenna. If you can see a tree anywhere on your lot, you've just found a home for your next antenna—and it will probably outperform any indoor design. I must admit, however, that I still stare out the window and dream of 100-foot towers and stacked Yagis!

CHAPTER NINE
RECEIVING ANTENNAS

By H. H. Beverage, ex-W2BML and Doug DeMaw, W1FB   From *QST*, January 1982

The Classic Beverage Antenna, Revisited

Established theory is timeless, but many amateurs do not have access to the archives that contain classical data of present-day interest. Medium-frequency DXers should appreciate this update on an historical 1922 *QST* article.[1]

W hy the Beverage or "wave" antenna? That's a question the seasoned 160-meter DXers need not ask, for many of them have used Beverage antennas to enhance the effective signal-to-noise ratio while attempting to extract weak signals from the sometimes high levels of atmospheric noise and QRM. Alternative antenna systems have been developed and used over the years, such as loops and long spans of unterminated wire on or slightly above the ground, but nothing seems to surpass the Beverage antenna for 160-meter weak-signal reception.

The practical limitation for many amateurs is the size of their property: a Beverage antenna must be a wavelength or greater in dimension, which for 160-meter work requires a minimum practical antenna length of 166.6 meters (546.8 feet) at 1.8 MHz (feet = meters × 3.281). In an ideal situation, one would deploy a number of Beverage antennas in order to facilitate weak-signal reception from a variety of favored directions, such as Europe, South America, Africa and Oceania. The magnitude of the property size requirements for

such a system might seem incomprehensible to the urban amateur, but the objective can be, and frequently is, realized by amateurs who live in rural areas. Some amateurs are part-time users of Beverage antennas. That is, they erect one or more of these antennas for short periods of time (with the kind permission of neighbors), mainly to improve reception during 160-meter contests and DX operations. One well known top-band DXer has for many years stretched a Beverage antenna across and beyond an interstate highway (not recommended) for use during 160-meter contest weekends.

The property requirements are complicated further by the need for an effective ground system at the terminated end of the wave antenna. Although the ground screen or radials are normally buried a few inches below the surface of the earth, one cannot, without permission, bury a ground system on someone else's property. Some amateurs have reported reasonable success by driving a number of rods into the ground near the terminating resistor, then bonding the rods to one another by means of heavy con-

ductive strap. However, the characteristics of the antenna are subject to change with the season in accordance with the conductivity of the soil, which is determined in part by the moisture content. The same is true, but to a lesser extent, when buried radials are employed for the ground screen.

Numerous attempts have been made to develop short or "baby Beverages," but the performance was always a compromise to that of a full-size Wave Antenna.[2] It is recognized, however, that some improvement in mf weak-signal reception is better than none, so the shortened version of a Beverage antenna may be worth investigation by those who have limited property.

It is ironic that arrays of small receiving loop antennas, operated in phase and simultaneously rotatable, have been proven to be highly effective in medium- and low-frequency weak-signal reception. But these arrays also require considerable property if they are to be utilized correctly. Furthermore, the cost of such a system, as opposed to a Beverage antenna, is substantially greater.

There seems to be a popular misconception about the frequencies for which the Beverage can provide the stated performance. It is not a suitable antenna for high-frequency reception. One must follow the general rule that applies to loop antennas: *employ the Beverage antenna at medium frequencies and lower.* Although some have reported improved reception from Beverage antennas at 3.5 and even 7.0 MHz, the suggested upper-frequency limit is 2.0 MHz. Occasional improved reception at hf may result from propagation conditions at a given time, but because the incoming sky waves above medium frequency arrive at moderate and high angles, and with changing polarity because of being reflected from the ionosphere, the Beverage is not suited to effective use in that part of the spectrum. The wave antenna is responsive mostly to incoming waves of low angle—those that tend to follow the contour of the earth and maintain a constant polarization. This reasoning is applicable to loop antennas as well. The apparent effectiveness of Beverage antennas above 2.0 MHz probably results from a reduction in local QRN and QRM off the sides and back of the antenna. A loop antenna would provide a similar improvement in reception, especially if a sense antenna were included to ensure a cardioid response.

The successful deployment of a Beverage antenna is dependent in part on understanding the concept and development of the system. The following text has been taken from the original disclosure in the amateur literature, which appeared under the H. H. Beverage byline in November 1922 *QST*. With the recent return of the 1.8 to 2.0-MHz band to U.S. amateurs, and with the easing of the earlier power restrictions in that band, it seems timely to present the original paper again.

Theory and Development

The Wave Antenna, which later became known as the Beverage Antenna, is a uni-directional antenna. It was developed by author H. H. Beverage, Chester Rice and E. W. Kellogg of the General Electric Co., and is covered by patents and applications. The Wave Antenna was first brought to the attention of the amateurs by Paul F. Godley, who described it in his report on the reception of American amateurs at Ardrossan, Scotland.

Theory

If a wire is suspended in space, it has a certain capacitance and inductance per unit length, which bear a definite relation to each other. This relation may be expressed as $1/\sqrt{LC} = V$, where V is a constant. This constant is the velocity of light. For example, if L and C are expressed as the capacitance and inductance per meter, then $V = 3 \times 10^8$ meters per second, which is the velocity of light. If a larger wire is used, or if two or more wires are used instead of one, in the ideal case the inductance decreases in the same ratio as the capacitance increases, so that $L \times C$ is always a constant. This means that, for the ideal wire, the currents induced in that wire will always travel along it at the velocity of light, independent of the size or number of wires.

A Beverage Antenna needs to be supported at several points and must run horizontally within a few feet of the earth. The effect of the supporting insulators and the proximity of the earth is to increase the capacitance in a greater ratio than the inductance decreases, so the velocity of the currents on a practical wire is always somewhat less than the velocity of light. On short wavelengths, however, the velocity approaches very close to the velocity of light, generally between the limits of 85% and 98% of the velocity of light for 200 meters (1.5 MHz), depending upon the size and number of wires.

In Fig. 1 is shown the simplest form of Wave Antenna. It consists simply of a wire, at least one wavelength long, stretched in the direction of the transmitting station. For explanation purposes, it may be assumed that the transmitting station is east of the receiving station, and that the receiver is placed at the west end of the antenna, as shown. The traveling wave from the transmitting station moves from east toward the west at the velocity of light. As the wave moves along the antenna, it induces currents in the wire that travel in both directions. The current that travels east moves against the motion of the wave and builds down to practically zero if the antenna is one wavelength long. The currents that travel west, however, travel along the wire with practically the velocity of light, and, therefore, move along with the wave in space. The current increments all add up in phase at the west end, producing a strong signal as shown by curve A in Fig. 2. In a like manner, static or interference originating in the west will build up to a maximum

at the east end of the antenna as shown by curve B in Fig. 2.

If the east end of the antenna were open or grounded through zero resistance, all of the energy represented by curve B would be reflected and would travel back over the antenna to the west end, where part of the energy would pass to earth through the receiver and part would be reflected again, depending upon the impedance of the receiver input circuit. The horizontal plane

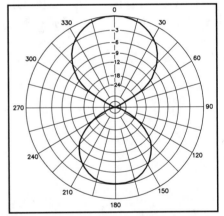

Fig 3—Directivity pattern of a Beverage antenna that is one wavelength long. It does not have a damping impedance included.

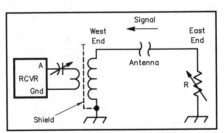

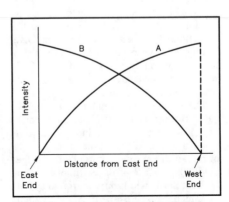

Fig 1—The simplest form of Wave Antenna.

Fig 2—Curve A shows how the current increments add in phase at the west end of the antenna. Curve B illustrates how the static and interference add at the east end of the antenna (see text).

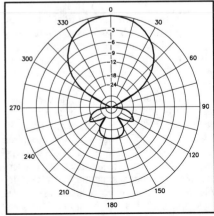

Fig 4—Directivity pattern for the antenna of Fig 3. The antenna has been damped properly.

intensity diagram would be bidirectional, as shown in Fig. 3. The reception from the west is not as good as from the east because some of the energy is lost because of attenuation in the wire as the reflected wave travels back from east to west

In order to make the antenna unidirectional, it is necessary to stop the reflections at the end farthest from the receiver end. This is accomplished simply by placing a noninductive resistance between the antenna and ground at the far end. If this resistance is made equal to the surge impedance of the wire, it absorbs all of the energy and prevents any of it from being reflected back to the receiver. The intensity characteristic becomes unidirectional, as shown in Fig. 4.

The value of the surge impedance depends upon the size, number and height of the wires above ground, but is independent of the length of the wire. For practical construction with one or two no. 12 copper wires, the surge impedance lies between 200 and 400 ohms. The surge impedance is theoretically equal to $R = \sqrt{L/C}$; where L and C are the inductance and capacitance per unit length.

Godley used the simple form of wave antenna shown in Fig. 1. However, this is not the most practical form, as it is necessary to go to the far end to make adjustments of the damping resistance.

Feedback Antenna

If two parallel wires are used, the Wave Antenna becomes very flexible, and the receiver may be placed at either end with local control of the damping. In Fig. 5, for reception from the east, the receiver at the west end is replaced by the primary, P, of transformer T2. The primary is coupled to the secondary, S, as closely as possible, and feeds the energy over the two wires as a transmission line. A second transformer, T1, at the east end, feeds the energy from the transmission line into the receiver. The energy fed over the transmission line circulates around the line as in an ordinary telephone line and, therefore, the currents pass through both halves of the primary of T1 in the same direction, inducing voltages in the secondary, that feed into the receiver. On the other hand, currents coming over the wires as an antenna, that is, from the west, are equal and in phase on both wires, and upon passing to ground through the two halves of the primary of the output transformer, T1, they pass through the winding in opposite directions and neutralize. With this circuit, the energy reaching the receiver is the same as it would be if the receiver were placed at the west end, except for the transmission-line losses, which ordinarily are 20 to 25% with proper design. With this feedback system the operator can make adjustments of the surge resistance without leaving the station, and can listen to the signals while he or she is making the adjustments.

Fig. 6 is equivalent electrically to Fig. 5, but in this case T2 has been replaced by a simpler circuit. By grounding one wire and

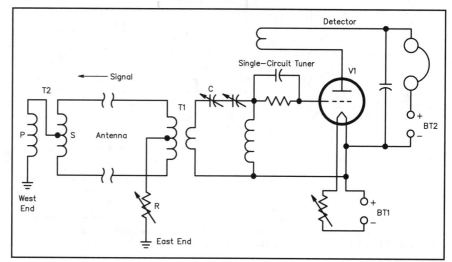

Fig 5—The receiver at the west end of the antenna is replaced here by primary P of T2. (See text.)

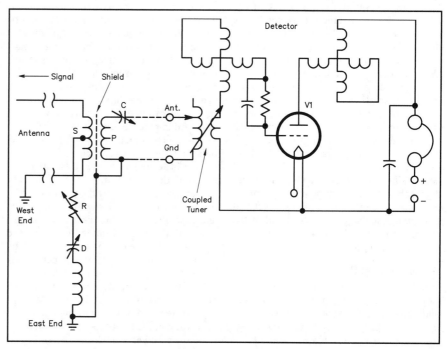

Fig 6—This circuit is equivalent to that of Fig 5, except that T2 has been replaced by a simpler circuit. The damping circuit is labeled "D".

leaving the other wire open, the energy is reflected on each wire, but the reflected currents on the transmission line are 180 degrees out of phase on the two wires and, therefore, a difference of potential exists across the terminals of the primary of T1, exactly the same as when the reflection transformer, T2, of Fig. 5 was used. If the ground resistance at the reflecting end is zero, the reflection of energy with the connections of Fig. 6 would be 100% efficient, and the only loss would be the transmission-line losses. The open ground reflection connection is preferable to a transformer, on short wavelengths particularly.

It is possible to damp a two-wire antenna from either end. In the case of Fig. 6, the signal from the east built up to a maximum at the west end, and was then reflected up to the east end, where the receiver and

damping circuit were placed. In the case shown in Fig. 7, the receiver is placed at the west end as in the case of the simple antenna of Fig. 1. Instead of placing the damping circuit at the east end, however, it is placed across the transmission line at the west end, where the receiver is. This damping circuit is practically just as effective as it would be if actually placed at the far end. This circuit also has the advantage that the desired signals do not pass over the transmission line, and the transmission-line losses are avoided.

In order for the damping circuit to be effective, it is necessary that the two wires of the antenna be joined through an inductance that is of high impedance compared with the impedance of the damping circuit. The best way to accomplish this result is to use a coil with a midpoint tap, as shown at

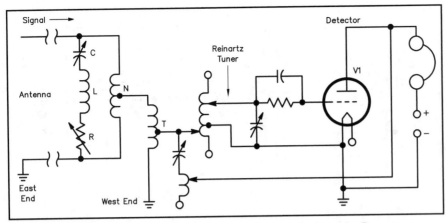

Fig 7—This example shows the damping circuit, D, across the two-wire Beverage antenna. The value of the damping resistance will vary with the wavelength.

N in Fig. 7. With respect to the transmission line, the two halves of this coil are adding, so the inductance across the line is high. With respect to the receiver, however, the two halves of the coil are opposing, so that the impedance in series with the output transformer amounts only to the leakage reactance of the coil, N, which can be made very small. A satisfactory inductor for N, for 200 meters, was a 24-turn coil, 7 inches in diameter, with a tap at 12 turns for feeding the output transformer, T. This coil was about 0.3 mH across the line, or 1900 ohms at 300 meters (1 MHz), and nearly 3000 ohms at 200 meters, which was high enough to have no appreciable influence on the damping circuit, and yet had low enough leakage reactance to allow the signals to pass to the receiver without noticeable weakening.

Damping Circuits

In Figs. 6 and 7, damping circuits D are shown that consist of resistance, inductance and capacitance in series. Because of distortion on the antenna, to back-wave effects, to interfering signals or static coming from such a direction as to be received on one of the little "ears" on the back of the antenna, as shown in Fig. 4, and so on, it often happens that there are appreciable residuals that are desirable to eliminate. This is possible by making the damping-circuit reactance either slightly capacitive or slightly inductive, instead of purely resistive. In some cases it may be desirable to reflect a small amount of energy to neutralize undesirable signals from the back end. This is readily accomplished by adjusting the resistance and capacitance of the damping circuit. The capacitance and inductance in this damping circuit are usually found to practically neutralize each other for the best adjustment; that is, they should tune approximately throughout the band of wavelengths it is desired to receive. If the wavelength being received is varied over wide limits, it is necessary to readjust the damping circuit capacitor for best results, although the adjustment is usually quite broad. The resistance does not need

readjustment except in special cases.

For a range of 180 to 360 meters (1.66 to 0.83 MHz), the damping circuit consists of an inductance of about 0.08 mH, a variable capacitor of 0.0015-μF maximum capacitance and a noninductive variable resistance in steps of 1 ohm from 0 to 500 ohms. A decade box is ideal for this purpose. However, ordinary wirewound potentiometers (inductively wound) have been used with success in damping circuits. It is necessary to select a potentiometer with sufficiently low inductance to tune well below the shortest wave it is desired to receive; then the inductance of the potentiometer is taken into account when calculating the value of inductance to be used in series with the resistance and capacitance. In this manner the inductance of the potentiometer used for the variable resistance may be tuned out, and the damping circuit may be made a pure resistance for any one particular wavelength.

When the damping circuit is placed across the transmission line as shown in Fig. 7, the value of the damping resistance may vary considerably with wavelength, becoming lower for short wavelengths, owing to the increase in attenuation at short wavelengths partially damping the antenna. In other words, the transmission line acts as a resistance in series with the damping circuit, and the transmission-line resistance becomes appreciable at short wavelengths.

Antenna Design

It is obvious from the theory of the Wave Antenna just given that it must point toward the desired signals or directly away from the desired signals. In case the antenna is pointed away from the signal, then the maximum signal occurs at the far end and must be brought up over the transmission line to the receiver, as shown in Fig. 6. In case the antenna is pointed toward the signal, it is necessary to put the damping circuit on the transmission line, as shown in Fig. 7. It is possible to use a single antenna for reception from either direction by switching arrangements to change to either the connection of Fig. 6, or that of Fig. 7,

at will. It is preferable on short wavelengths to point the antenna toward the signal, using the connections of Fig. 7, but the feedback of Fig. 6 gives practically the same results except that the signals are not quite as loud as a result of the transmission-line losses.

It is necessary to run the Wave Antenna in as straight a line as possible and not nearer than 200 feet (61 m) to other parallel wires, such as telephone and power lines, as the influence of these wires is liable to distort the directive characteristic of the antenna. Other wire lines may be crossed at right angles without undesirable effects. In cases where it is not feasible to run the Wave Antenna in line with the desired signals, it is possible to get good reception with the antenna somewhat "off line" by sacrificing signal intensity. By referring to Fig. 4 it is seen that for the average antenna one wavelength long, it is possible to be 45 degrees off line before the signal drops to half intensity. Beyond 45 degrees the signal falls off very rapidly. Twenty degrees off line, the signal intensity has fallen off only 10%, so very good reception may be obtained. If the antenna is two wavelengths long, it is more directive, and it is not possible to receive well if it is more than 25 or 30 degrees off line.

The antennas are constructed of copper or other nonmagnetic material, although Cutler of W7IY reported in October 1922 QST that he had obtained good results on a galvanized-iron wire. The size of the wire is usually between no. 10 and no. 14 B&S, although it is possible to get fair results even with no. 18 bell wire. The usual construction is to put up two wires on a cross arm about 2 to 3 feet long. The wires are suspended by porcelain cleats, or in more permanent construction standard telephone pins and high-grade insulators are used.

The height of the wires above ground has a marked influence on the velocity of the currents along the wires when the wires are close to the ground, but if the wires are 10 feet above the ground there is little to be gained in velocity by making them higher, as shown in the curves of Fig. 8. These data were taken on an antenna at Belmar, New Jersey, by H. O. Peterson. This antenna extended over fairly conducting soil. The character of the soil underneath the antenna influences the velocity to some extent, but the data of Fig. 8 are about the average velocity. These curves show that the velocity becomes lower at longer wavelengths.

If the velocity is too slow, then the currents in the wire lag in phase behind the wave in space, and a point is soon reached when the current in the wire from the far end is so far behind in phase that it not only does not add to the increments from points close to the receiver, but may actually subtract. The maximum length that it is feasible to use is that length at which the current in the wire lags 90 degrees behind the wave in space. This length is given by

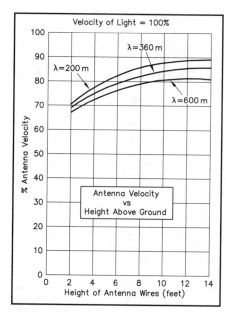

Fig 8—Curves that show the antenna velocity factor as a function of the height above ground.

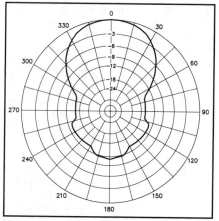

Fig 9—Directivity pattern of a Beverage antenna with an effective height of 15 meters, with a vertical or end effect of 3 meters superimposed upon it.

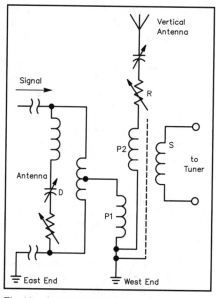

Fig 10—A separate vertical antenna can be used to balance out the end effects discussed in the text. The circuit arrangement is depicted here.

$$L = \frac{\lambda}{4\left(\frac{100}{C} - 1\right)}$$

where

λ = wavelength in meters

C = signal velocity on antenna expressed in percent velocity of light.

For example, from Fig. 8 we find that the velocity of the currents in the two wires suspended at a height of 10 feet is about 88% of the velocity of light for 200 meters, so the maximum usable length is:

$$L = \frac{200}{4\left(\frac{100}{88} - 1\right)} = \frac{200}{0.544} = 367 \text{ meters}$$

Therefore, it is not feasible to use a two-wire antenna suspended at a height of 10 feet for an antenna that is more than two wavelengths long for 200 meters. By increasing the height, the velocity will increase, and longer wires may be used. Fig. 8 shows that the velocity increases slowly with the height about 10 feet, so the wires must be much higher to be of material advantage. Making the wires too high introduces a difficulty on short waves that does not occur on long waves, and that is the "end" or vertical-antenna effect. The effective height of a 200-meter Wave Antenna is about 5% to 10% of its horizontal length, depending upon the nature of the earth beneath the antenna, and so on. If an antenna is 200 meters long, therefore, the effective height will be between 10 and 20 meters. If the antenna is on supports 10 feet high, the vertical or end effect may be equivalent to an effective height of nearly 3 meters (10 feet), distorting the directive curve. In Fig. 9 is shown the directive curve of a Wave Antenna of 15-meters effective height with a vertical or end effect of 3 meters superimposed upon it. It will be noted that the end effect may mount up to very serious proportions if the antenna is made too high. It is, however, possible to balance this end effect by means of a separate vertical antenna, as shown in Fig. 10. P1 is the standard primary, while P2 is a second primary coil of about the same number of turns, which is wound over P1 but in the opposite direction. In practice, however, the end effects seem to be very much smaller than predicted theoretically, so as a general rule if the antenna is not over 10 feet high the end effects are so small that it is not worth the trouble to balance them. From the foregoing considerations, it is evident that 10 feet is a good average height for short Wave Antennas.

Design of Transformers

With the feedback circuit of Fig. 6, only one transformer is necessary. The output transformer, T1, was wound on a 7-inch cardboard tube. The primary, P, was 20 turns of no. 24 enameled copper wire, with a tap at 10 turns or the exact center. Over the primary was placed a shield consisting of a piece of tinfoil insulated from both windings by means of paper. This shield was grounded to cut out capacitive currents between primary and secondary. It is important that the tinfoil or other metal foil be not quite a complete turn around the primary; the ends must not touch or it will act as a short-circuited turn and introduce high losses. The secondary consisted of 5 turns of no. 18 bell wire wound over the tinfoil shield. The center of the secondary winding was lined up carefully over the center of the primary winding; otherwise the transformer would not be balanced. With the circuit of Fig. 6, the transformer balance was tested by opening both wires at the west or reflection end. When T1 was properly balanced, the receiver was quiet, indicating that the two halves of the primary were perfectly symmetrical with respect to the secondary. T1 of Fig. 6 was designed to work with a coupled receiver. The secondary of the output transformer was connected in series with the primary winding of the receiver input transformer and was tuned by the series capacitor, C. For 200 meters, it is usually better to use a separate capacitor, C, outside of the tuner capacitor as shown in Fig. 5, but for longer wavelengths this series capacitor may be omitted.

When the circuit of Fig. 7 was used, the transformer just described was used with success, but better results were obtained by cutting the primary turns down to 15 instead of 20. This transformer is shown in Fig. 1, but may be used with the connections of Fig. 7. A metal-foil shield is used between primary and secondary, and is grounded as shown. In all of these transformers the coupling between primary and secondary should be as close as possible.

Fig. 7 illustrates an auto-transformer, T. The total turns are 15, and the receiver is tapped off at 5 turns. The diameter of the turns is 7 inches, but smaller diameters have been used by increasing the number of turns to obtain the same inductance. This auto-transformer connection was once adapted to a Reinartz tuner with excellent results by Roland Bourne, W1ANA, at W2BML.[3]

Surge Resistance and Velocity Factor

The velocity factor and surge resistance were easily determined by oscillator tests. An oscillator was coupled to the antenna, as shown in Fig. 11. A coupling coil, L, was included in the antenna circuit. It consisted of only two turns. The far end of the antenna was left open for the first test, and a resonance curve of the antenna was taken. The curve is plotted as curve A in Fig. 12. Then both wires of the antenna were grounded at the far end, and the resonance curve taken again. This is shown as curve B in Fig. 12. In order to find the velocity, it is necessary to calculate what the resonance points would be if the velocity of the currents on the wires were equal to the velocity of light.

The length of the antenna was carefully measured. In the case of this particular antenna at Belmar, New Jersey, the length was 240 meters. Assuming that the velocity of the currents on the antenna is equal to the velocity of light, the first resonance point with the far end of the antenna open will be the quarter-wave oscillation, as in an ordinary antenna. The wavelength will be $4 \times 240 = 960$ meters. The next resonance point will be the three-quarter-wave oscillation, or $^4/_3 \times 240 = 320$ meters. The next will be the $^5/_4$ oscillation, or $^4/_5 \times 240 = 192$ meters, and so forth, for all odd multiples of the quarter-wave oscillation. In a like manner, with the far end of the antenna grounded, the antenna will oscillate at all even multiples of the quarter-wave oscillation. These calculated values are recorded in Table 1. In the next column, the observed values taken from Fig. 12 are recorded. By dividing the calculated value by the observed value, we get the actual velocity at that particular wavelength in terms of percent of velocity of light.

To determine the surge resistance, a noninductive resistance was placed between antenna and ground at the far end, and the resonance curve was taken again. Fig. 13 shows the results of this test on the Belmar antenna. Curve A with 500 ohms at the far end, shows broad but unmistakable resonance points at open oscillation wavelengths. On the other hand, curve B with 200 ohms at the far end, shows grounded resonance points. Curve C, with 300 ohms at the far end, shows no resonance points, indicating that the antenna is quite aperiodic. Therefore the surge resistance for this particular antenna is approximately 300 ohms. The downward bend of curve C below 200 meters is not caused by the antenna, but results from the oscillator output falling off when the coupling capacitor approached zero setting.

When one of the wires was grounded at the far end, the other wire was left open and the damping resistance was placed

across the wires at the station end, as shown in Fig. 7, a smooth curve, similar to the curve C of Fig. 13, was obtained when the noninductive resistance was 500 ohms. In this case, however, there were slight irregularities in the curve that do not appear in curve C of Fig. 13.

Fig. 14 shows the resonance and damping curves taken on a single-wire antenna by R. B. Bourne at W2BML/W2EH. This wire was 195 meters long, and was suspended from trees at a height varying from 15 to 20 feet. It was interesting to note that Bourne's antenna had a velocity of approximately 93% of the velocity of light at 200 meters and, therefore, showed that a single wire could be used up to a length of over three wavelengths, or approximately 2000 feet. Such an antenna should also show very directional properties, but lacks the flexibility and ease of adjustment of the two-wire antenna.

Performance

Two 200-meter Wave Antennas were erected at Belmar, one running west from the station and the other running south. These antennas were arranged with switching such that the connections of Fig. 6 or Fig. 7 could be selected at will on either antenna. That is, the west antenna could be used for reception from either the east or the west, and the south antenna could be used for reception from either the north or south. For comparative purposes a flat-topped single-wire antenna, 40 feet high, was erected. The effective height of this vertical antenna was estimated as approximately 8 meters. The signals on the Wave Antennas were about 50% stronger than on the vertical, giving an effective height for the Wave Antennas of 12 meters. This figure corresponds to about 5½% of the horizontal length of the Wave Antennas.

Listening tests on these antennas showed marked directive properties, as expected. Listening south, most of the stations heard were in the third and fourth districts, but careful adjustments were necessary to eliminate second-district stations to the north. With the antenna directive toward the north, the best reception was from the first and second districts, although several eighth-district stations were heard. The east-west antenna worked better than the north-south antenna, probably because the ground resistance at both ends was less than an ohm, whereas the ground resistance at the far end of the north-south antenna was very high (nearly 300 ohms) making it difficult to operate the damping circuit effectively. The reception from the west was excellent, great numbers of Midwest, Southwest and West Coast cw stations being heard without interference from first and second-district stations. With the antenna directed east, only local W2s, Long Island W2s and a few W1s were heard. There was considerable QRN reduction at times on the eastward reception, as the QRN was often heavy in the south or west.

On the 360-meter broadcast station wavelength, very good results were experienced in eliminating interference, particularly when using the antenna for west reception and cutting out New York and Schenectady interference. Station WOC at Davenport, Iowa, was received particularly well on the Wave Antenna at times when reception was impossible on the vertical antenna, owing to local interference.

Even on 600 meters, these Wave Antennas showed very good directivity, particularly for reception from ships at sea.

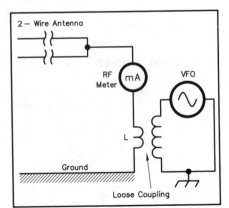

Fig 11—An oscillator can be coupled to the antenna, as shown here, to determine the velocity factor and surge resistance.

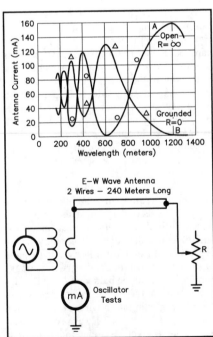

Fig 12—Curves obtained with oscillator tests of a 240-meter-long Beverage antenna (see text).

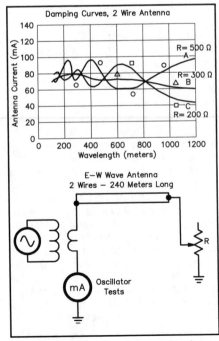

Fig 13—Damping curves for a two-wire Beverage antenna (see text).

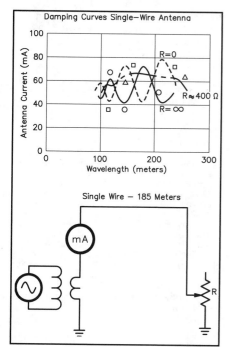

Fig 14—Damping curves for a single-wire Beverage antenna.

Table 1
Calculation of Velocity of Currents on Antenna
Length—240 meters, 2 no. 10 wires, 3 meters high

Mode of Oscillation	Wavelength Calculated	Wavelength Observed	Vel/Wires Vel/Light
$^1/_4$	960	1200	80%
$^2/_4$	480	590	81%
$^3/_4$	320	390	82%
$^4/_4$	240	280	86%
$^5/_4$	192	220	87%
$^6/_4$	160	180	89%

Bourne's antenna at Riverhead, Long Island, ran in a direction about 10 degrees north of west. He reported his results as follows: "Signals from the south and southwest come in with about 25% to 50% increase in signal strength over a vertical antenna 60 feet high. Signals from New England are, in general, very weak, and in some cases cannot be heard at all when using the Wave Antenna. No interference from ships or shore stations using commercial wavelengths was noticed. WSA, at Easthampton about 20 miles away, at times had a very strong harmonic on about 225 meters, which interfered seriously with 200-meter reception when the ordinary antenna was used, but because this station was southeast, no interference was experienced when using the Wave Antenna. Radiophones on 360 meters came in with about the same intensity as with the vertical antenna, but often the signal-static ratio was much improved with the Wave An-

tenna, and, as with 200-meter reception, interference from WSA and WBC (East Moriches, 10 miles away) was entirely done away with."

The amount of static reduction experienced with the 200-meter Wave Antenna at Belmar depended entirely upon the distribution of the static at different times. On several occasions a marked improvement was noted in the signal-static ratio when receiving from the east and north, and sometimes when receiving from the west, but it was rarely observed to make any marked improvement when receiving from the south.

The author wishes to acknowledge the valuable assistance received from Messrs. H. O. Peterson, R. B. Bourne and A. B. Moulton, in the collection of these data on the 200-meter Wave Antennas.

Practical Considerations

The foregoing text from the 1922 *QST* article discusses slight differences in overall performance with respect to the wire gauge used in a wave antenna, with the smaller-diameter wire being the less desirable choice. In a practical amateur installation it is unlikely that one could discern a performance difference without having two antennas to compare—one with heavy-gauge wire and one with, say, no. 20 wire. Many amateurs have reported good results when using the smaller wire sizes for single-wire Beverage antennas. But, if the heavier wire is available, it should be employed in the interest of optimum performance. The longevity of the system under the stresses of wind and icing will be superior when the antenna is made from no. 10 through no. 16 wire. If for some reason it is desired to have a measure of "invisibility" for the antenna, one should not overlook the possibility of using light-gauge wire.

Quality insulators are required at the support points of the wire. Some amateurs have merely secured their Beverage antennas along the span by wrapping the conductor around tree trunks and fence posts. This is not recommended if proper performance is desired. The incoming signal energy should be able to traverse the wire without propagation discontinuities and losses along the antenna length. Good insulators will help to make this possible.

The least complex of the Beverage antennas is the single-wire version, although

the two-wire type offers greater flexibility of adjustment. In any event, the integrity of the termination and ground system is a matter of prime importance. Some amateurs have simply driven an 8-foot pipe into the ground at the far end of the antenna, then attached the terminating resistor to it. Depending upon the earth conductivity at a given location, this technique may represent no ground system whatsoever! A quality ground system contains a substantial number of buried radial wires, as is the case with quarter-wavelength vertical antennas.

If an extensive ground arrangement isn't practical, the amateur should use as much wire as possible, even if some of the radials are quite short. Sufficient wire should be used to ensure that the ground resistance is as low as possible.

Other Considerations

Fig. 15 illustrates a Beverage antenna used by W1FB (then W8HHS) in Michigan for 160-meter reception in the early 1950s. It was roughly 1500 feet (three wavelengths) long, which posed no physi-

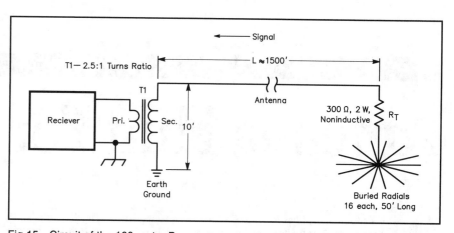

Fig 15—Circuit of the 160-meter Beverage antenna used at W8HHS for DX reception. The primary and secondary windings of T1 were returned to separate ground points to resolve unwanted common-mode bc-band signal coupling that affected receiver performance.

cal problems on the 40-acre farm site. The terminated end was toward the northeast to accommodate reception from Europe. The transmitting antenna was a 60-foot vertical with center loading and 20 buried radials that were dispersed uniformly from the base of the vertical to a length of 80 feet each. Signals that could not be heard in the noise while receiving with the vertical could be elevated above the atmospheric noise by as much as two S units when using the Beverage antenna for DX work to Europe. Owing to the majority of the noise fronts existing to the southwest, in the Gulf of Mexico region, and because of the back-rejection of the Beverage, such an improvement was possible. Heavy QRN could often be heard with the vertical, even though the weather was clear locally and for a thousand miles or more to the southwest. Noise from storms can be propagated a great distance when conditions are otherwise good at 1.8 MHz.

One problem that was experienced with the antenna of Fig. 15 became manifest as severe receiver overloading from a nearby commercial a-m station on 1240 kHz. The receiver dynamic range and front-end selectivity of that period were generally anything but spectacular. Hence, cross-modulation and other overload effects were not uncommon to 160-meter operators. The difficulty was resolved by breaking up the common-mode transfer path from the antenna to the receiver. At first the return ends of the primary and secondary windings of Tl were brought a common ground point. By returning the low end of the Tl primary to the receiver ground terminal and the low end of the secondary to the earth ground, the overloading ceased. The Beverage antenna was an effective collector of bc-band energy! T1 was used to provide a broad-band transformation from 300 ohms (unbalanced) to 50 ohms unbalanced at the receiver input. A small TV-set flyback trans-

former-core was used in the transformer. A 900 μ$_i$ ferrite toroid core would be excellent for the purpose today.

References

[1]H. H. Beverage, "A Wave Antenna for 200-Meter Reception," QST, November 1922, p. 7. The professional disclosure of the wave antenna was presented by H. Beverage, C. Rice and E. Kellogg ("The Wave Antenna, a New Type of Highly Directive Array"), in the Transactions of the AIEE for 1923. It contains 51 pages of technical information.

This presentation of the original QST work by Beverage has been edited for style, tense and terminology to bring it up to present-day QST technical language. The diagrams, curves and radiation patterns have been re-drafted to conform to present-day symbology and style. Nothing else has been changed. The reprint of the article is presented in smaller type size to differentiate between the writing of H. H. Beverage and D. DeMaw.

[2]B. Booth, W9UCW, "Weak Signal Reception at 160—Some Antenna Notes," June 1977 QST.

[3]J. Reinartz, W1QP, "Some Further Improvements in My Tuner," QST, October 1922, p. 12.

Antenna Hint For DXers

A word of caution on special low-noise receiving antennas—put them as far away from your vertical or semi-vertical transmit-ting antenna as possible. The distance should be one wavelength or more. If there is insufficient separation, the receiving antenna will pick up the same noise and crud you've tried to avoid. The noise can be reradiated from the transmitting antenna. — *Stewart Perry, W1BB*

By Floyd Koontz, WA2WVL　　　　　　　　　　　　**From *QST*, February 1995**

Is this EWE for You?

Here's a really simple receiving antenna for 80 and 160-meter DXing.

As we approach the bottom of the sunspot cycle, MF and HF signals are weak and ambient noise is high. What every serious DX operator needs to improve signal-to-noise ratio is a directive receive antenna. In a previous *QST* article,[1] I described an array of tuned loops. That array has excellent front-to-back (F/B) directivity, but it is quite narrowband and requires careful construction and alignment to get full performance.

A New Design

One fundamental of directive wire-antenna design is that two or more parallel wires carrying similar RF currents produce directivity regardless of the phase or amplitude of the currents, if the wires are separated significantly (greater than 0.05 λ). I surmised that if a short wire (less than ¹/₄ λ) were formed as an upside-down **U** and fed against ground, good F/B directivity would result.

Using Brian (K6STI) Beezley's *AO* software, I modeled such an antenna (with a terminating resistor to ground at the far end

[1]Floyd Koontz, WA2WVL, "A High-Directivity Receiving Antenna for 3.8 MHz," *QST,* Aug 1993, pp 31-34.

of the wire). Success was immediate, with F/B ratios greater than 30 dB predicted. Since *AO* assumes that the ground beneath the antenna is perfectly conducting, the initial result was suspect.

I loaded the *AO* dimensions into K6STI's *NEC-WIRES* software, which takes into account the ground conductivity and dielectric constant beneath the antenna. Compared to *AO's* output, the results dove into the 15 to 20-dB region, but the F/B was real. With some readjustment of the antenna dimensions and the value of the far-end resistor, the F/B increased to more than 30 dB—that from a piece of wire between two ground stakes! What could be simpler?

A Beverage It Ain't

This antenna (Figure 1) looks amazingly like a Beverage, but its method of operation is entirely different. By analyzing the *RUN* file in *AO,* you can see that for optimized height and length, a phase shift of about 45° occurs between the center of wire 1 and the center of wire 3.

Taking into account the 180° phase shift caused by the direction of the currents in wires 1 and 3, the total phase shift is about 135°, with wire 3 leading (a reflector). From this, you can see that the Ewe receiving antenna is essentially a two-element driven array in which the rear element (wire 3) is fed via the top wire (wire 2). Wire 3 has about 65% as much current as wire 1. Figure 2 shows the azimuth and elevation patterns of a single 80-meter Ewe over average ground. Because part of the antenna is horizontal, I was concerned about noise pickup degrading the overall results. The horizontal gain computed to be about 20 dB lower than the vertical gain, and at a high angle off the side.

One surprise is that the F/B holds over several megahertz. That led me to design a second antenna covering 1.8 to 4.0 MHz. The calculated F/B is greater than 25 dB over this entire range without adjustment. This design will be shown later.

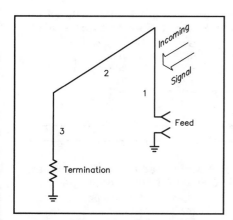

Figure 1—The Ewe is the receiving antenna. The antenna end attached to the resistor is the rear. The antenna produces a cardioid pattern.

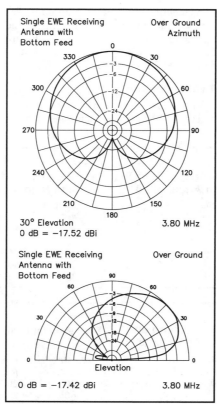

Single EWE Receiving Antenna with Bottom Feed　　　　Over Ground Azimuth

30° Elevation
0 dB = −17.52 dBi　　　　3.80 MHz

Single EWE Receiving Antenna with Bottom Feed　　　　Over Ground

0 dB = −17.42 dBi　　　　3.80 MHz

Figure 2—Azimuth and elevation patterns of a single Ewe over average ground.

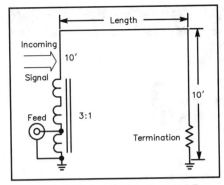

Figure 3—A 10-foot-high bottom-fed Ewe with a 3:1 impedance-matching transformer and a resistive termination.

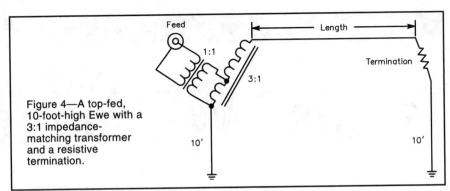

Figure 4—A top-fed, 10-foot-high Ewe with a 3:1 impedance-matching transformer and a resistive termination.

Feeding and Terminating the Ewe

Figures 3 and 4 show two ways of feeding the Ewe. The basic feed impedance is about 450 Ω, so a transformer with a turns ratio of 3:1 provides a good match. The value of the terminating resistor varies according to the wire height, wire length and band of operation, and whether bottom feed or top feed is used. I devised the top-feed system so that the RG-58A coax could be run overhead. Because I considered a wire 10 feet high something that could be managed easily by most amateurs, I used *NEC-WIRES* to generate Tables 1 through 4 using a 10-foot height as one parameter.

Reversible Ewe for 80 Meters

In order to experiment with this antenna, the first one built had identical 3:1 turns ratio ferrite transformers (50 to 450 Ω) at both ends of the Ewe. Two RG-58A coax cables ran to the shack so that the feed and the termination were at the receiver. This allowed nulling real signals and noise arriving by skywave to determine the antenna's effectiveness. I quickly discovered that a high F/B could not be obtained with a purely resistive termination, so my old Beverage control box was pressed into service. This control box has an adjustable RLC termination, which was just what I

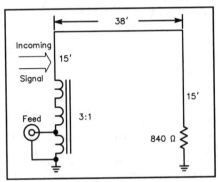

Figure 5—A 160/80-meter Ewe 15 feet high.

needed. Results on nulling skywave signals ranged from minimal to as high as 40 dB. This is as expected, since only stations directly off the back are in the pattern null. The termination consists of a 500-pF air-dielectric variable capacitor, a 4.7-μH inductor (a low-Q inductor is adequate), and a 100-Ω potentiometer, all in series.

160 and 80-Meter Design

As mentioned before, a compromise design was made to cover 1.8 to 4 MHz. The optimum dimensions are shown in Figure 5. Gain at 1.8 MHz calculated as –22 dBi and at 4 MHz, –12.3 dBi.

Two-In-Line Array

Next, I investigated an array of two Ewe antennas. When two elements are fed as an

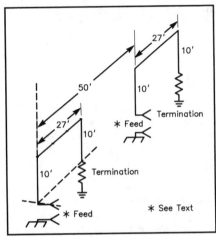

Figure 6—An array of two 10-foot-high Ewes.

endfire array, the forward beamwidth remains the same, but the pattern off the back improves. Figure 6 shows the in-line physical configuration that gives excellent results on 80 meters. Feeding both antennas with equal currents and 135° phasing provides the patterns shown in Figure 7.

Two-in-Broadside Array

Two Ewe elements, side by side, can be fed in phase to produce a beam that is much narrower in azimuth. Element spacing should not exceed ⅝ λ (about 162 feet at 3.8 MHz). Figure 8 shows the physical configuration

---

Table 1
Ewe Parameters for Bottom Feed at 3.8 MHz, with an Antenna Height of 10 Feet and Elevation Angle of 30°

Ant. Length (feet)	F/B (dB)	Gain (dBi)	Termination (Ω)
15	17.9	–21.4	650
20	23.5	–19.5	740
25	34.8	–18.0	850
30	27.0	–16.9	870
35	20.1	–15.9	900

Table 2
Ewe Parameters for Bottom Feed at 1.8 MHz, with an Antenna Height of 10 Feet and Elevation Angle of 30°

Ant. Length (feet)	F/B (dB)	Gain (dBi)	Termination (Ω)
30	23.3	–27.2	1000
40	35.2	–25.1	1060
50	28.6	–23.4	1200
60	18.2	–22.1	1220

Table 3
Ewe Parameters for Top Feed at 3.8 MHz , with an Antenna Height of 10 Feet and Elevation Angle of 30°

Ant. Length (feet)	F/B (dB)	Gain (dBi)	Termination (Ω)
15	15.9	–22.1	825
20	19.3	–20.0	910
25	24.9	–18.3	960
30	40.4	–17.0	970
35	26.6	–16.0	980

Table 4
Ewe Parameters for Bottom Feed at 1.8 MHz, with an Antenna Height of 10 Feet and Elevation Angle of 30°

Ant. Length (feet)	F/B (dB)	Gain (dBi)	Termination (Ω)
30	20.3	–27.5	1150
40	29.8	–25.1	1230
50	30.3	–23.4	1290
60	20.8	–22.0	1300

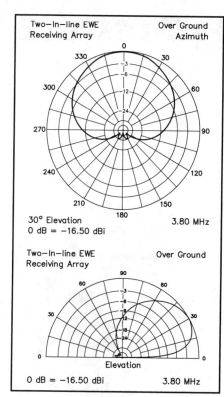

Figure 7—Azimuth and elevation patterns of the dual-Ewe array of Figure 6.

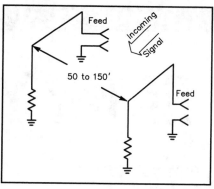

Figure 8—A pair of side-by-side Ewes fed in phase to produce a narrower azimuth pattern (see Figure 9).

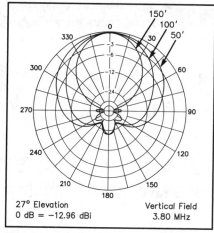

Figure 9—Azimuth patterns of the Figure 8 array for varying degrees of antenna separation.

used, and Figure 9 shows how the pattern varies with element separation. Use equal-length feed lines run to a common central point, beyond which the lines can be placed in parallel.

Construction

Little needs to be said about how to build this antenna. Height and length are not critical and small ropes or cords can be used to pull up the top wire. I stretched my experimental antenna from the corner of my house to a convenient wooden pole. Of course, trees or towers will do fine as supports. I recommend that the upright wires be not closer than 10 feet to a tall metal tower and that the antenna not receive in the direction of a supporting tower.

For the ground rods, I used 2-foot, 8-inch pieces of copper-plated steel, made by cutting 8-foot ground rods into three pieces, but common pipe will work. The

small ferrite transformer and the potentiometer are mounted in small diecast boxes equipped with binding posts on the top. The boxes are clamped to the ground rods using stainless-steel hose clamps. Any type of copper wire can be used for the antenna.

I use a pair of 11-dB low-noise amplifiers to bring signal levels up to levels comparable to those provided by full-size antennas.

Summary

Antennas much simpler than this—that play as well as this—are hard to come by. Give a Ewe a try. I'm sure you'll find it improves ewer receiving capability on the low bands!

More EWEs for You

You've seen the basic EWE, now you can have switchable, directional EWEs!

S ince the introduction of the EWE antenna in *QST*,[1] many amateurs have improved their low-band receiving capability with a EWE. In the correspondence I've received was a letter from Tony Kazmakites, WB2P, who suggested switching wires above the feedpoint to change directions. This article uses that idea and deals with two switchable configurations of the EWE providing coverage in all directions.

Single EWE for Four Directions

Figure 1 shows a EWE designed to receive in any one of four directions. Relay switching at the base of the central support allows selection of any one of the four wires. As you may remember from the previous article, the *back* of a EWE is the *terminated end,* so the relay box is always at the *front* of the antenna. You select the *southwest wire* to receive from the *northeast.*

All four wires are anchored to a central square wooden pole and are brought to the bottom of the pole on separate faces (see Figurse 2 and 3). The wire opposite the relay box is brought through a small hole near the base of the pole. Wires from the top of the pole are run at 90° from each other, although other angles (or more wires) can be used (see Figure 3). In this design, the terminations are at the bottoms of the outer poles. A 2-kΩ potentiometer mounted in a small diecast aluminum box (Bud CU-123) allowed experimentation, but a fixed-value resistor can be used instead. Figures 4 and 5 show the azimuth and elevation patterns on the single EWE optimized for 160 meters (modeled with the other three antennas in place).

Construction

I'll describe the construction that I used to build this antenna in an open field. You can stretch the wires out in nearby woods, or tie them to existing support points. All five supports are made of 12-foot-long, 4 × 4-inch pressure-treated timbers. Each pole is set two feet into the ground and anchored using one or two bags of ready-mix concrete (80 to 160 lb). My soil is hard clay. If your antenna farm has sandy or loose soil, a more substantial base may be required. The five poles are oriented so that their

sides are parallel to each other. Screw eyes are installed near the top of each pole and about one foot above the ground on each pole, one screw eye on each pole face. There's one set of screw eyes on each of the outer poles and four sets of screw eyes on the center pole.

I drove 4-foot-long ground rods into the earth near the lower eyes of the outer poles and planted one rod near the relay box on the center pole. In soft soil, longer rods would be appropriate. (If you can easily drive the rod into the ground, it's not long enough!) Some EWE-antenna builders who have rocky soil and can't use ground rods have reported good success using an above-ground return wire (counterpoise). Because the feedpoint impedance of this antenna is usually above 400 Ω and the terminating impedance is generally above

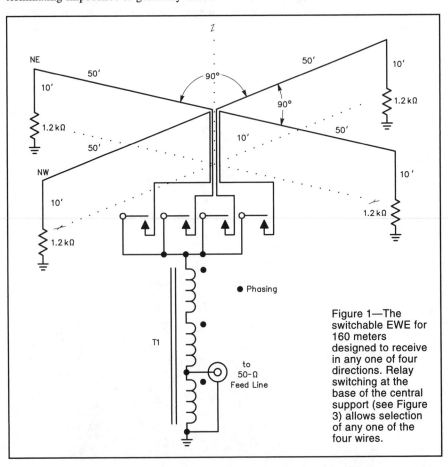

Figure 1—The switchable EWE for 160 meters designed to receive in any one of four directions. Relay switching at the base of the central support (see Figure 3) allows selection of any one of the four wires.

Figure 2—The center pole of the EWE array. All four antenna wires pass through large eyebolts and descend to the bottom of the pole on separate faces 90° apart from each other. Other angles (and/or more wires) can be used.

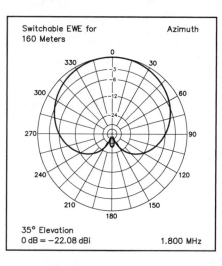

Figure 4—Azimuth pattern of the 160-meter switchable EWE.

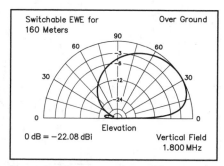

Figure 5—Elevation pattern of the 160-meter switchable EWE.

Figure 3—Near the base of the center pole, an aluminum enclosure houses the four relays. The four wires from the top of the pole are anchored at the pole base in large eye-bolts, one on each face of the pole. Each wire's bottom end connects to the inside of the relay box through binding posts. From the hidden pole face, the wire passes through a hole drilled in the pole near the eyebolt above the relay box. The feed line approaches the pole through a piece of PVC pipe and is attached to the box via a BNC connector. A ground rod connected to the relay box finishes the installation.

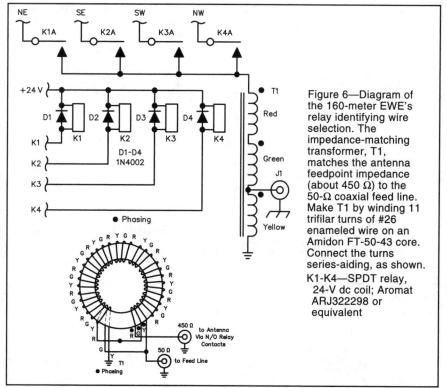

Figure 6—Diagram of the 160-meter EWE's relay identifying wire selection. The impedance-matching transformer, T1, matches the antenna feedpoint impedance (about 450 Ω) to the 50-Ω coaxial feed line. Make T1 by winding 11 trifilar turns of #26 enameled wire on an Amidon FT-50-43 core. Connect the turns series-aiding, as shown.
K1-K4—SPDT relay, 24-V dc coil; Aromat ARJ322298 or equivalent

800 Ω, you can discount the effect of ground resistance.

Relay Box

Four relays are housed in a small diecast aluminum box (Bud CU-234) mounted six inches above ground level (see Figure 3). The relays are mounted on the box *lid* along with the connectors and four insulated binding posts. In building an assembly such as this, I line the inside of the box lid with a piece of $^1/_{32}$-inch-thick, double-sided, cop-per-clad PC-board. Such material is thin enough to work with BNC bulkhead jacks and pro-vides a solderable ground every-where. The ground-rod connection is made to the box, which remains fastened to the post should you want to remove the relay assembly for maintenance. Figure 6 shows the relay antenna-switching arrangement. In the future, I plan to control the relays via the coaxial-cable feed line using a serial data stream (I have a 500-foot feed-line run).

Many readers of the first EWE article had questions about the transformer used to match the feedpoint to the feeder. I use a small (about $^1/_2$-inch diameter) core with a permeability of 850 (such as an Amidon FT-50-43) wound with 11 trifilar turns of #26 enameled wire. The turns are connected series-aiding as shown in Figure 6. If you don't have access to trifilar wire, simply combine three individual strands of enameled wire.[2]

Tuning the EWE

Several EWE builders have attempted to vary the termination resistance to obtain a good impedance match at the input. I *don't* recommend this because maximum F/B *doesn't occur* coincidentally with the termination value that gives the best match! The EWEs described in this article are terminated with 1.2-kΩ resistors and may display an SWR of up to 2.5. Receiving antennas need *not* be matched to low SWR values unless they are part of a matched, phased array. Adjust the termination resistance value *only while receiving a signal from the rear of the antenna.*

Two-Element Array

Later, I placed a second group of five poles 300 feet northeast of the previously described antennas. Figure 7 shows an arrangement of eight EWEs that gives excel-

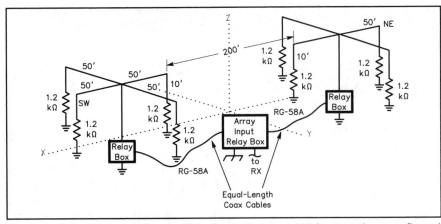

Figure 7—Here's how two sets of switchable EWEs (about 300 feet apart) are configured to form a broadside/endfire array. This arrangement of eight EWEs provides excellent directivity in any of four directions. For an endfire pattern, one element is fed through a 180° delay line. For a broadside pattern, the two selected elements are fed in phase.

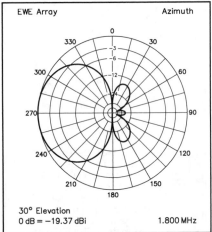

Figure 8—Azimuth pattern of the endfire array.

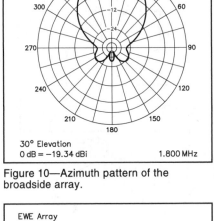

Figure 10—Azimuth pattern of the broadside array.

Figure 9—Elevation pattern of the endfire array.

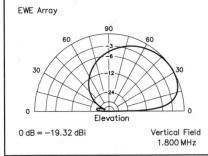

Figure 11—Elevation pattern of the broadside array.

Table 1
Relay States

| | Relays Energized | | |
Direction	Group 1	Group 2	Array Input Box
NE	K1	K1	K2
SE	K2	K2	—
NW	K3	K3	K1
SW	K4	K4	—

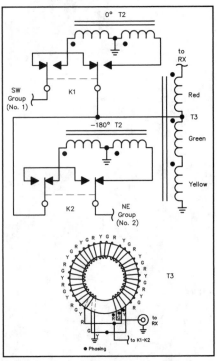

Figure 12—Diagram of the phase-switching box used for selecting northeast endfire, southwest endfire, or the broadside mode of the array of Figure 7. The 180° delay for the endfire mode is produced using a bifilar-wound transformer, T2, (with the center tap grounded). One transformer is required for each direction. T3 is made like T1, but the feedpoint of T3 is tapped up two *windings* from the ground end. The relays are shown with K2 energized to produce a northeast endfire pattern. The relay states required for reception in each of the four directions are shown in Table 1.

summer, this antenna found only a small amount of use to date. This winter, however, I expect my EWEs to greatly increase my 160-meter operating pleasure!

I've shown you two ways to use the EWE element to provide receiving directivity in four or more directions on 160 meters. I'm sure "EWE" will discover many other ways of configuring the EWE. Have fun!

Notes

[1] Floyd Koontz, WA2WVL, "Is This EWE for You?" *QST*, Feb 1995, pages 31 to 33.
[2] Use wires of different colors to make identification easy or spot color-code the wires using nail polish or spray paint of different colors. For more information on building various transformers, see Jerry Sevick, W2FMI, *Transmission Line Transformers* (Newington: ARRL, 1990), 2nd edition. This publication is out of print and no longer available from the ARRL Bookshelf, but a copy may be available at your local library, ham club or from a friend. See also Joseph J. Carr, K4IPV, "Make your own RF Transformers," *Popular Electronics,* Nov 1995, pages 67 to 69 and 93.—Ed.

lent directivity in any of four directions. To the northeast or the southwest, an endfire pattern is produced. A broadside pattern is available to the southeast or northwest. Figures 8 and 9 show the endfire patterns; Figures 10 and 11 show the broadside patterns. For an endfire pattern, the front element is fed through a transformer (T2) to achieve a 180° delay. For a broadside pattern, the two selected elements are fed in phase.

Array-Relay Switching

Figure 12 shows the switching arrangement I use for selecting northeast endfire,

southwest endfire, or the broadside mode. The 180° delay for the endfire mode is produced using a bifilar-wound transformer, T2, (with the center tap grounded). One T2 is required for each direction. T3 has the same basic design of T1, but the feedpoint is tapped up two *windings* from the ground end. The relays are shown with K2 energized to produce a northeast endfire pattern. The relay states required for reception in each of the four directions are shown in Table 1.

Summary

With band conditions as they are in the

By Edwin A. Andress, W6KUT

From *QST*, September 1995

A K6STI Low-Noise Receiving Antenna for 80 and 160 Meters

Here's how you can make a low-cost, low-noise, 21-foot-square receiving antenna that's certain to boost your DXability.

In his accompanying *QST* article,[1] Brian Beezley, K6STI, brings to our attention the details of how an antenna can be designed to eliminate (or greatly reduce) power-line noise to aid low-band reception. Brian's article includes the description of a 25-foot-square receiving antenna and matching section along with some observed operational characteristics. While Brian explored antenna design trade-offs with his computer software program, with his guidance, I built several prototype antennas. Here, I'll share the performance, operational and construction information I gained as Brian progressed to final design.

Although I had to shorten the sides of the antenna to 21 feet to fit my available real estate, the results achieved at my location in omnidirectional signal reception and power-line-noise cancellation with such a simple 10-foot-high antenna are remarkable! This deviation illustrates the fact that you can make substantial dimensional changes without performance degradation provided symmetry is maintained and the resulting feedpoint impedance is matched to the feed line.

The higher the Square, the greater its output level. After using a 21-foot Square on 160 meters, Brian and I are confident that small Squares can be made for each band and perform well. The antennas can be easily mounted on towers using horizontally positioned poles (such as quad antenna spreader poles[2]), or installed on posts as phased arrays to get directivity in a small area.

The Test Location

The Square is an ideal antenna for hams living on small lots located near power lines that generate noise heard on the 40, 80 and 160-meter bands. My location is ideal for testing. It is between—and about one block from—the arms of a **Y** formed by a 30-year-old, 6.9-kV ac San Diego Gas and Electric distribution system that has had its voltage increased without benefit of insulator upgrade, its load increased by changing to higher power-handling transformers without benefit of increased wire size, and its maintenance service decreased as the power company strives to reduce cost: a typical suburban area suffering from gradual hardware degradation that leads to line-noise generation. This area is located 12 miles from the Pacific Ocean and is subject to considerable day-to-day variations of temperature and humidity. During the test period, there was plenty of rain and the overall noise level was lower than normal, many times only S7 rather than the usual S9+. However, we still had multiple noise sources to deal with while observing Square performance.

My lot lacks sufficient space to install even a modest Beverage, but its pie shape permits the installation of two 80-meter $1/4$-λ wave phased verticals, $1/4$ λ apart, at the rear of the property. The verticals (with a gain of 4.9 dB) work well for transmitting, but when receiving, all signals—however strong—become victims of great batches of power-line noise. The noise sources are multiple and surround the location. Stations frequently can be heard, but not copied well—or at all.

The Search for Quiet

Using a shortened 80-meter horizontal trap dipole at 84 feet and vertical rotary loops has not solved my noise problem. My two-element, full-size, 40-meter Yagi at 73 feet provides the best S/N, but the signals are very weak and my linear amplifier doesn't like it when I forget to throw the antenna switch before transmitting! It kicks off line and during the three-minute wait for it to return on line, I've usually lost the intended contact! I've been continually trying to improve reception with antennas, and have spent hundreds of dollars on black boxes. Both have helped on single noise sources, but they've been unable to cope with multiple noise sources.

The EWE Antenna

A EWE antenna described by Floyd Koontz, WA2WVL, in *QST* [3] was undergoing tests when discussion of my results with Brian triggered his research that led to the Square design. I found the EWE to be a good antenna. I aligned its cardioid patterns on long path and short path, the same paths as my two $1/4$-λ verticals. The vertical component of my line-noise sources, however, was still overwhelming when picked up by the two vertical portions of the EWE, leaving me with an unacceptable S/N. After the EWE, I tried a horizontal, 25-foot-per-side triangular antenna, 10 feet high. Noise was reduced from that picked up by the EWE, but was still more than I had hoped for. At this point, Brian decided on a square antenna design to get complete noise cancellation between both halves of the antenna. I then built the 21-foot Square.

Scouting for Noise Sources

Once the antenna was ready, Brian visited my radio shack and, following a quick check, decided that we needed a survey of noise generation on the premises. I chased around turning off various devices to eliminate noise received on various antennas, or through the house wiring: We found several contributors. As a result, we recommend that you frequently survey your home and immediate surroundings to identify and eliminate noise from fluorescent light fixtures, variable voltage controls for lights, touch-controlled lamps, bug-killer traps, electric blankets, electronic air-purifying systems, packet terminals, computers and so on.

Overall Performance Observations

Our survey showed us that reception of noise from ac noise sources within a household may not be eliminated or reduced by the Square (such noise may be delivered on, or radiated from, the ac lines strung through the house), but that power line noise about one block away will be. At my location, the Square antenna pattern appears to be omnidirectional and effective in reducing noise from multiple power-line sources. Summer—when everything dries out and arcing is everywhere—will test its omnidirectional capabilities to the utmost.

During testing, an unexpected bonus of the Square surfaced: We noticed that the antenna frequently drops background noise several S units. It's our assumption that this noise is a collection of lower-level suburban-area ac noise, so the effect we experienced may only be apparent in cities and towns. Whatever the sources, the noise reduction is a boon to reception. Theory and observations say that atmospheric noise and electrical-storm disturbances (noise crashes) won't be eliminated by the Square, but my Timewave DSP-9 audio filter helps lower atmospheric background noise level, and a noise blanker helps on certain types of repetitive impulse noise—even some types of power-line noise. The combination of the DSP-9 and the Square is more effective than either one alone.

Performance

After a month of testing, I find that listening to SSB and CW signals on 80 and 160 meters is incredibly more comfortable using the Square, and I've been able to copy weak DX signals on the Square that I could hear—but not copy—using any of the other antennas. (An extract from my log is shown in Table 1.) Although my 160-meter performance checks were not as exhaustive as those I performed on 80 meters, I compared my DX reception with the local 160-meter Big Guns daily and found that I could copy all the signals they heard, but for a shorter time during the window. I'm so encouraged with the results that I'm now erecting a vertical for transmitting.

It's important to realize that the Square is not a substitute for a directional antenna. If you can install long Beverages, do so. However, the omnidirectional nature of the Square may still come in handy in the absence of many Beverages if multiple noise sources exist.

Table 1
Performance Comparison of the Antennas at W6KUT

	Phased Verticals (no preamp)	Dipole at 84 ft (no preamp)	K6STI Square (20-dB preamp)
3814 kHz LSB*	9+	9+	8+
AM Noise	7.5	7.5	3
LSB Noise	6.5	5.5	1
CW Noise	6.5	5	0

S Meter Readings for

Compare the noise-only S-meter readings with those of the signal + noise readings.
Note: The S-meter readings were taken using a Yaesu FT-1000D equipped with a meter movement linear between S4 to S9.

*LSB signals from W5 area stations were being received on this frequency.

Noise *only* for the various modes was measured on a clear frequency (3783 kHz).
Filter bandwidths: AM, 6 kHz; LSB, 2.4 kHz; CW, 500 Hz.

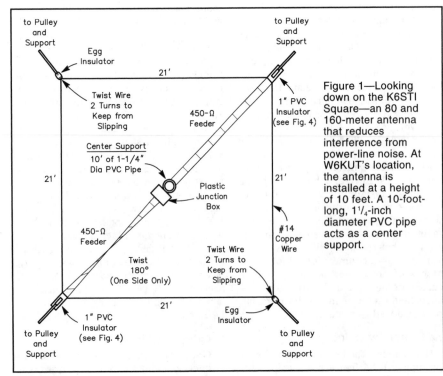

Figure 1—Looking down on the K6STI Square—an 80 and 160-meter antenna that reduces interference from power-line noise. At W6KUT's location, the antenna is installed at a height of 10 feet. A 10-foot-long, 1¼-inch diameter PVC pipe acts as a center support.

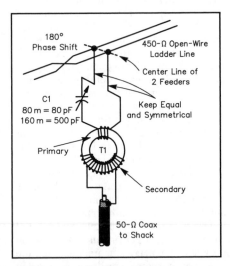

Figure 2—Details of the matching network for 80 and 160-meter single-band operation when using 450-Ω open-wire ladder line. This network is included inside the junction box (see Figure 1) of each 80 and 160-meter antenna. (A Radio Shack 270-223 plastic box makes a suitable enclosure). C1 is a compression trimmer or air variable capacitor with a capacitance of 80 pF for 80 meters and 500 pF for 160 meters. C1 is connected in series with one leg of T1's primary winding. T1 consists of an FB-77-1024 ferrite shielding bead wound with #24 enameled wire. For 80 and 160 meters, the primary has 4 turns. The secondary winding consists of 10 turns for 80 meters and 20 turns for 160 meters.

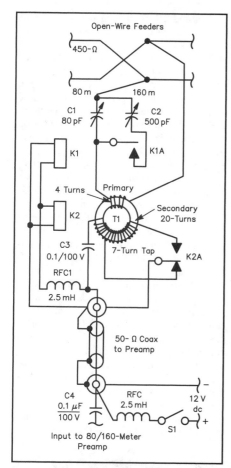

Figure 3—Matching network for a duo-band (80 and 160-meter) K6STI Square. Relays, powered through the coaxial-cable feed line, provide the required LC switching. A Radio Shack 270-223 plastic box is used for the enclosure.

C1—80-pF air variable capacitor or compression trimmer

C2— 500-pF air variable capacitor or compression trimmer

C3, C4—0.1-μF, 100-V disc-ceramic capacitor

K1—SPST relay, 12-V dc coil (Radio Shack 275-135)

K2—SPDT relay, 12-V dc coil (Radio Shack 275-241)

RFC1, RFC2—2.5 mH, 100 mA

S1—SPST toggle switch

T1—Amidon FB-77-1024 shielding bead wound with #24 enameled wire; 4-turn primary, 20-turn secondary, with a tap at the 7th turn.

Construction

The Square is simple to build, install and tune. Figure 1 shows the Square as viewed from above. Here I'll tell you what's required to make 21-foot Squares for 80 and 160 meters, or a single Square that works on 80 or 160 meters at the flick of a switch.

Each Square is 21 feet on a side, 10 feet high and made of inexpensive #14 TW house wire. The antenna is fed at opposite corners using feeders made of 450-Ω open-wire ladder line made with plastic spacers positioned about 6 inches apart. The two feed-line sections are each about 15 feet 6 inches long. Length isn't critical, but the sections must be of *equal* length to maintain antenna symmetry. Some sag in the

lines is acceptable if they are cut long, but I prefer to cut the sections as exactly as possible and then pull the supporting lines taut to reduce wind whipping. Be sure to keep the antenna as square and level as possible in a horizontal plane even though the ground beneath the antenna slopes away.

At the Square's center, the two pieces of ladder line join inside a 4 × 6-inch plastic box and attach to the 4-turn primary winding of a matching transformer (T1) via a loop-resonating capacitor (see Figure 2). T1's secondary winding connects to an SO-239 connector from which 50 feet of RG-213 run to the shack. C1 can be placed at the middle of the primary winding to maintain symmetry. For ease of construction, however, I placed the capacitor in series with one side of the primary winding with no detectable performance degradation.

For C1, I use a surplus compression trimmer that tunes above and below the resonance point. On 160 meters, a ³/₄-inch-square, multiplate, mica-dielectric compression capacitor is used; its original capacitance range ran from 475 to 1800 pF. Calculations showed a need for about 475 pF, so I removed plates until there were only three per side and ended up with a capacitance range of 250 to 780 pF.

T1 consists of a primary and secondary winding of #24 enameled wire wound on an Amidon FB-77-1024 core. If your antenna has dimensions different from those shown—or if the feed-line section impedance is other than 450 Ω—you may have to experiment with the tap placements on T1 to find a match. If the dimensions and

feed line are about the same as shown, the given turns numbers should work.

Figure 3 illustrates how the feed system for my antenna is matched for both 80 and 160 meters. An SPST switch and a 2.5-mH RF choke provide relay operating voltage via the feed line. This permits remote selection of 80 and 160-meter operation. At the antenna end, dc flows through another RF choke. C3 and C4 are 0.1-μF blocking capacitors used to protect the input of the preamplifier or receiver from the relay operating voltage and to keep from shorting the dc to ground in the junction box.

Two relays are used to permit best symmetry on the feed-line side where additional capacitance is added for 160-meter operation using an SPST relay. Selection of a 7-turn tap for 80 meters or a full 20 turns for 160 meters requires an SPDT relay. (Note: If an 80-meter transformer only is made, use four turns on the antenna side and 10 turns on the feed-line side). Small, 12-V dc relays are available from many surplus houses and from Radio Shack for less than $2 each.

Figure 4 shows an inexpensive, removable and adjustable support pole and homemade PVC insulators used to hang the Square. The junction box containing the matching network is supported by a 10-foot length of 2¹/₄-inch schedule 40 PVC pipe. A reducer and an 18-inch-long detachable ³/₄-inch-diameter PVC extension are attached to the box. The pole is set 18 inches into the ground at the center of the Square. The removable extension permits you to build the antenna at ground level, then raise

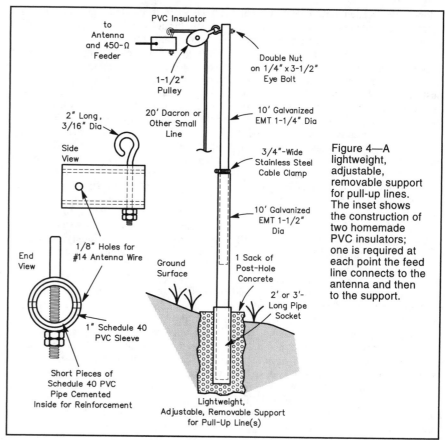

Figure 4—A lightweight, adjustable, removable support for pull-up lines. The inset shows the construction of two homemade PVC insulators; one is required at each point the feed line connects to the antenna and then to the support.

or lower it as needed with the pulleys. The Dacron line I use to raise and lower the antenna has proved itself reliable during years of service and I recommend it.

When assembling the Square, remember that *one* of the open-wire feed lines must be rotated 180° to obtain power-line noise cancellation by the two halves. To get good noise cancellation, maintaining symmetry in the two halves of the antenna is important. Maintain symmetry in each antenna leg, from the start of the two feed-line sections to the center junction, and from the junction to the matching transformer winding inside.

Tuning the Square

We used an MFJ-208 HF/VHF SWR analyzer and a Startek 35-BG frequency counter to tune the antenna, by feeding output of the analyzer to the coax connector or feed line from the shack. (Any of the presently available antenna analyzers should do the job.) Using the given values, all that should be necessary is to sweep the frequency range and find the meter dip. (Use an insulated tuning wand to eliminate hand-capacitance effects when adjusting compression trimmer capacitors.) The frequencies should bracket 3650 kHz for 80 meters. If they don't, use a different value of capacitance. Set the oscillator to 3650 kHz and adjust the capacitor to get a minimum SWR; 1:1 should be possible within the capacitor's range. If an SWR of 1:1 isn't achievable, T1's turns ratio may require adjustment. A turns ratio change should be required only if a major change is made in antenna or feed-line geometry.

I don't recommend using a transceiver for tuning the antenna because the matching network capacitor's dielectric may be punctured. If you do use your transceiver, keep the output power at its lowest level.

However, we found that because of the low power output from the MFJ, AEA and other antenna analyzers, the reflected power reading on 160 meters can be misleading where significant fundamental, or harmonic, broadcast-station RF is present. This energy is detected and indicates as reflected power; the indicator registers full scale and nulls aren't visible. At K6TQ, with his 135-foot-tall, 6-inch-diameter vertical, a nearby broadcast station on 1000 kHz overwhelmed the MFJ and AEA instruments, so we resorted to using a transceiver. At my location, no broadcast signals interfered with the Square during testing.

For the two-band antenna, first tune the 80-meter section as described previously, then activate the relays (applying battery voltage locally) and adjust C1 for a 1:1 SWR at 1870 kHz.

Preamplifier

I pressed into service a homemade, all-band, three-pole band-pass filter used in multi-multi contest operations. The filter enclosure also contains a 20-dB preamplifier. I prefer a gain of 20 dB to keep levels within a reasonable range when switching between antennas. I have not found the band-pass filters to be of any help on receiving with the Square at my location. W6YA, however, installed an 80-meter band-pass filter to solve local AM broadcast station overload when using his horizontal 80-meter dipole.

Selecting Feed-Line Material

Our first tests were conducted using currently available 450-Ω feed line manufactured with alternating 1½-inch open

spaces and 1½-inch-long plastic sections (50% coverage). During our evaluation, we found that by setting the Square's resonance at 3650 kHz, we got S-meter readings at 3500 and 3800 kHz that were only about two S units down from resonance peak, and, with minor receiver-gain adjustment, there was no noticeable change during operation. We also found that resonance would drop about 200 kHz when the feed line was wet from heavy condensation or rain (San Diego never has ice, so that condition wasn't tested!). The 200-kHz excursion seemed a bit excessive, so we tried exterior 300-Ω TV line installed in a section of ³/₄-inch PVC tubing between the center and antenna feed-line connection points. This proved mechanically ungainly, but worse, matching-network requirements became very difficult to achieve with available components, so we abandoned the use of 300-Ω line.

Knowing that a large impedance variation exists from one end of the feed lines to the center, Brian dug out some old ladder line. It has small plastic spacers 6 inches apart and a conductor spacing of about 1 inch. We calculated the impedance of Brian's open ladder line to be 500 Ω, and two pieces of this line were installed without support, the antenna ends being pulled taut. Resonance shifted only from 3650 to 3495 kHz from dry to soaking wet, an acceptable shift. On

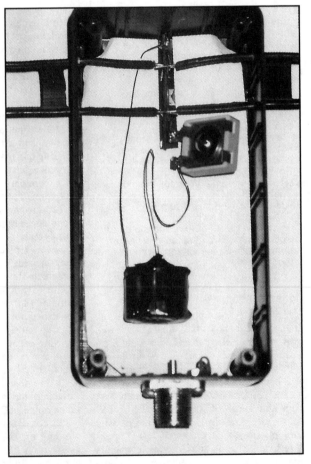

Figure 6—An inside view of the matching network enclosure. The secondary winding has not yet been added to the transformer (wrapped in electrical tape) and no connection yet exists to the SO-239 feed-line connector.

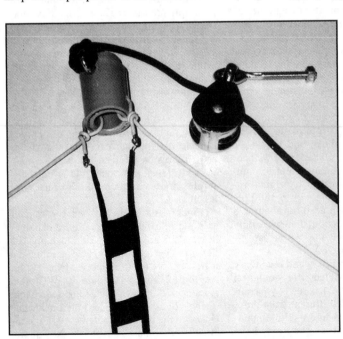

Figure 5—Attachment of the 450-Ω line to the antenna arms and homemade PVC support insulator.

A Search for Better Hearing

Although I live in the country, miles from any town, signal reception still suffers from local power-line noise. This noise generally manifests itself in two ways: First, as a single-source noise that has an identifiable direction and is most prevalent during damp foggy weather; second, as multiple sources generated by hot, windy weather conditions common much of the time in this area of California. Signal levels from either of these noise sources often are S9 or stronger. (Yes, I've had the local electrical utility company working on this problem—with little success.)

These noise levels greatly reduce my operating enjoyment particularly on 75 and 160 meters where I must (as many others do) use vertically polarized arrays. At least on 40 meters and up, I have rotary Yagis and can minimize the noise somewhat by antenna positioning.

A couple of years ago, I decided to attack the hearing problem on 75 meters. First, I made both of my transmitting antennas (a bobtail curtain for north/south directions and a delta loop for east/west) more directional. That provided noise rejection from some directions. I did this by adding two more bobtail curtains in line and spaced ¼ λ apart. These are phase switchable for unidirections of either 20° or 200° with about 20 dB F/B ratio.

For east/west, I added another delta loop, spaced ⅛ λ from the first, phase switchable for unidirectional east or west, with a F/B of about 15 dB. These changes helped reception and also added a bit of forward transmitting gain. But I still needed more noise reduction, so I installed a 2-λ, two-wire Beverage switchable for either 45° or 225°. This helps reduce atmospheric noise, but does little toward reducing the power-line noise. Still looking for improvements, I built Charlie (W7XC) Michaels' version of the Null Steerer.* This system really works great on the single-source noise. I can typically null such noise from S9 down to S2 or S3, and can also often null QRM such as RTTY. Unfortunately, this setup doesn't work well when the hot, dry winds arrive and the multiple noise sources fire up.

Along Comes the Square

When Brian, K6STI, and Ed, W6KUT, offered me material describing their noise-reduction work, I was more than happy to try another approach to reducing the noise level. I built Ed's version of Brian's Square and although I didn't build it in the most ideal environment—because of convenience and expedience—it really does work!

Even though I have five acres for my antenna farm, I'm running out of space with six towers and several wire antennas already planted. So, I nested my Square among these antennas. In addition, the Square is situated about 20 feet from a chain-link fence and about 50 feet from a barn covered with aluminum siding. One corner of the Square is supported by one of my towers. Two corners of the Square are supported by existing guy wires and the fourth corner of the antenna is supported by a wood pole attached to the chain link fence. However, each corner does have at least six feet of antenna rope separating it from its metallic support.

Even though my installation of the Square is certainly less than ideal, the antenna works. Any and all line noises are reduced simultaneously from typically S9 to S1 or less. With a 15 to 20-dB preamplifier in the receive line, the desired signal is nearly the same level as when received on the transmitting antenna. For example, I recently noted a mini pileup on 3795 kHz. The noise level was sizzling at S9 on my transmitting antenna and I *could not hear* the DX station. When I switched to the Square, there was C21/ZL1AMO at S5 (with 15 dB of preamplifier gain) and zero noise! I worked a good one that morning that would not have been possible without the Square. So, even if the space available for this antenna is restricted, be assured it will work—and work well. Try to keep the antenna as square and symmetrical as possible.

I'm going to build a Square for 160 meters and am looking forward to next fall when I can test another Square's effectiveness on the weaker 75-m long-path signals.—*Keith A. Fowler, W6BCQ*

*Charles J. Michaels, W7XC, "The Null Steerer Revisited," *QST*, Jul 1994, pp 29-33.

---

160 meters, the resonance shift was about half that of 80 meters.

It's important to realize that a shift in resonance will only cause a variation in signal level, *not* a change in S/N. We can easily tolerate a bad-weather resonance shift resulting in some reduction of level, but not a loss of S/N. The resulting SWR change is meaningless since we're not transmitting with the antenna. A small loss of signal level is the only penalty, and it's easily recovered by a tweak of the audio control.

Summary

A leading supplier of wire and cable reports that quality control of available open-wire line of the type we chose to use is poor, and that the spacers tend to fall off easily. If such line is used, find some way of securing the spacers.[5] In the future, we may make our feed lines using #14 wire with a conductor spacing of 2 to 4 inches and homemade spacers. A change in matching transformer turns ratio and resonating capacitor values may be required. Brian's configurations have kept the size of horizontal loop and feed-line impedances such that only a capacitor is needed to resonate the loop.

Brian feels that if the Square is raised to 20 feet to improve signal strength, feed lines could come down at an angle to a 10-foot-center pole (so that tuning could be easily achieved) without seriously degrading performance. This has not been tested.

If anyone builds a Square antenna and finds it caked in ice one day, please let Brian or me know what happens to the Square's resonance! Have fun constructing—and pleasant listening!

Notes

[1]Brian Beezley, K6STI, "A Receiving Antenna That Rejects Local Noise," *QST*, Sep 95.
[2]For information on fiberglass spreader sources, see *The ARRL Antenna Book*.
[3]Floyd Koontz, WA2WVL, "Is this Ewe for You?", *QST*, Feb 1995, pp 31-32.

CHAPTER TEN

ANTENNA IDEAS FROM W1JF

By R. R. Schellenbach, W1JF   From *QST*, June 1982

Try the "TJ"

Whether TJ makes you think of Cameroon or Tokyo, Japan, the "DXpertise" of this antenna could help you snag the rare ones.

City dwellers and small-lot owners frequently complain, "No room for a good DX antenna." Can you work DX on 160, 80 and 40 meters from that restricted bit of real estate? The answer is yes. Let me tell you about a compact antenna that is useful for working DX.

The TJ is a five-band, vertically polarized antenna system. In the 160-meter band the TJ is essentially a $1/4$-wavelength (λ) T (see Fig. 1). It becomes a $1/2$-λ T on 80 meters and a $5/8$-λ T on 40 meters. For 20 and 15 the configuration becomes a $1/2$-λ inverted J. It is from this combination of T and J that the antenna gets its name.

High performance is realized with the TJ on 80 through 15 meters because the maximum current point is elevated above ground. On 160 meters, the performance approaches that of a full-size, $1/4$-λ vertical antenna. The horizontal section of the TJ does not radiate appreciably. The current on each side is of equal magnitude and opposite phase, thus canceling radiation.

The three lower frequency bands employ a combination of top-loading techniques to physically shorten the antenna. The end sections act as capacitance hats on 80 and

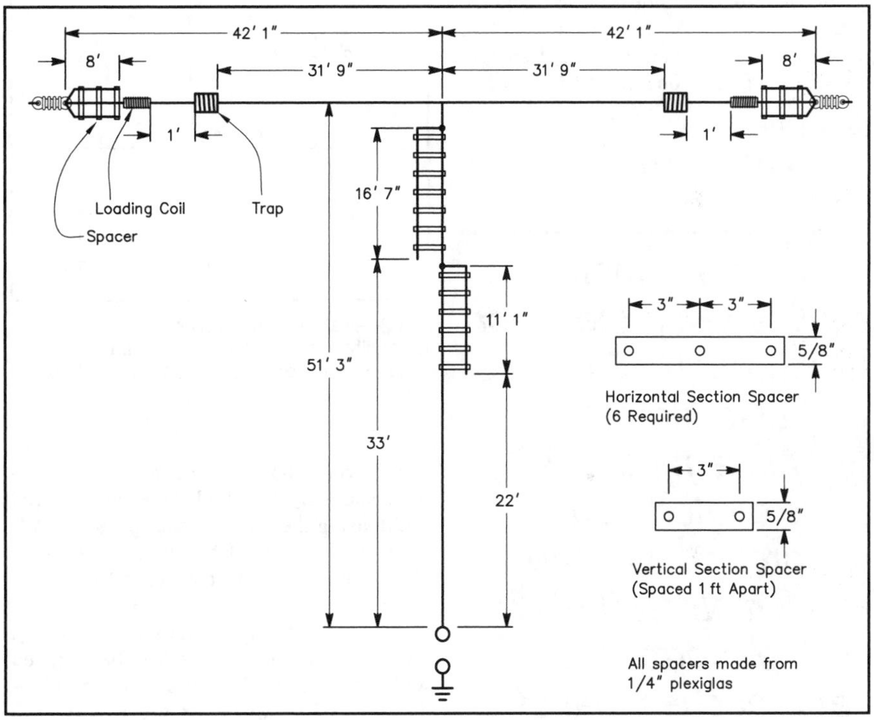

Fig 1—The TJ antenna.

Ground-wave coverage is very good, thanks to vertical polarization and a low angle of radiation. Best of all, that low angle accounts for the excellent DX results I have obtained while using the TJ.

160 meters. On those two bands there is an almost $2/3$ size reduction in the TJ. Because top loading is employed, bandwidth is not reduced as drastically as it would be if other methods were used. On 160 meters, top loading means a more desirable current distribution and a more favorable feed point impedance (30 to 40 Ω compared with 8 to 10 Ω for a base-loaded vertical). If the ground system has a resistance of 5 Ω, the TJ should be about 85% efficient on 160 meters. A base-loaded vertical would exhibit only half that efficiency. Better efficiency means more effective radiated power—exactly what we all want.

Construction Details

The loading coils are wound on 8-inch

lengths of $1^1/2$ in. PVC tubing.[1] Use 120 turns of close-wound, no. 14 enam. copper wire. I installed a pair of egg insulators inside each coil for support.

The 40-meter traps employ the same type of tubing and support scheme. They were constructed after a *QST* article by Johns.[2] I used RG-59/U coaxial cable and found resonance at 7.05 MHz, using 11 turns.

The dimensions shown in Fig. 1 were derived empirically. You can copy the measurements or modify them for operation in your favorite parts of the bands. I find resonance in my antenna at 1.815, 3.6, 7.05, 14.1 and 21.1 MHz.

Install the TJ in the clear, as far from surrounding objects as is possible. High

Table 1
Antenna Feed-Point Impedance

Band	Impedance (approx.)
160	35 Ω
80	>1000 Ω
40	100 Ω
20	>1000 Ω
15	>1000 Ω

quality glass or ceramic insulators should be used at the antenna ends. Nylon rope can be used to support the antenna. The feed point should be no more than 2 feet above the ground.

A good ground is required for efficient operation on 160 and 40 meters. My ground system covers 2 acres and employs a buried network of over 5000 feet of solid copper ribbon. You may not want to duplicate that, but you should install an effective ground system. Stanley described several possible configurations in *QST*.[3]

Tuning the TJ

An antenna-matching network is essen-

> *High performance is realized with the TJ on 80 through 15 meters because the maximum current point is elevated above ground. On 160 meters, the performance approaches that of a full-size, $1/4$-λ vertical antenna.*

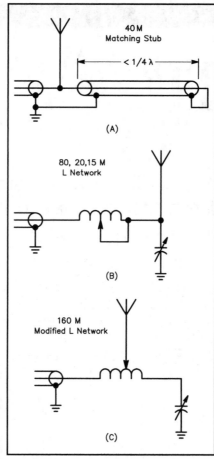

Fig 2—Matching networks for the TJ. Capacitance and inductance values should be determined experimentally for each band.

tial to proper operation of the TJ. The network should be installed at the antenna feed point, using the shortest leads possible. Adjustments can be set for the favorite band of operation, or you can do it by remote control.[4,5,6]

Feed-point impedances are given in Table 1. These impedances can be matched with the three configurations shown in Fig. 2. The exact values for these networks should be determined experimentally for each installation. Components for the matching networks should be mounted in a weatherproof housing.

Start with a quarter wavelength of co-

axial cable for the 40-meter matching stub. To find the length in feet, divide 234 by the frequency in megahertz and multiply by the velocity factor of the cable. (Velocity factor is 0.66 for polyethelyene dielectric and approximately 0.80 for foam.) Short the free end of the stub and observe the SWR. Now shorten the stub, short the end and check SWR. Continue this process until a satisfactory match is found. The stub can now be rolled into a coil and the end taped.

One nice feature of stub matching is bandwidth. As you move away from resonance, the reactance of the antenna and stub move in opposite directions. The reactances tend to cancel, thus providing greater bandwidth.

The matching circuit for 80, 20 and 15 meters is a simple L network. On 160 meters I employ a modified L network. I found that the best match and highest antenna current was obtained with the tap a little more than half way toward the variable capacitor. The TJ covers the entire cw portion of any of the five bands, with one setting of the antenna-matching unit.

Performance

Short-skip performance is not as effective as it is with a low horizontal antenna. Lack of high-angle radiation explains that characteristic. Ground-wave coverage is very good, thanks to vertical polarization and a low angle of radiation. Best of all, that low angle accounts for the excellent DX results I have obtained while using the TJ.

I found it satisfying and a lot of fun to build my own antenna. You would, too. Why not construct your own TJ? Good luck and good DX!

Notes

[1] mm = in. × 25.4, m = ft × 0.3048.
[2] R. H. Johns, "Coaxial Cable Antenna Traps," *QST* May 1981, p. 15.
[3] J. O. Stanley, "Optimum Ground Systems for Vertical Antennas," *QST*, December 1976, p. 13.
[4] H. Drake, Jr., "A Remotely Controlled Antenna Matching Network," *QST*, January 1980, p. 32.
[5] B. K. Imamura, "A T-Network Semi-Automatic Antenna Tuner," *QST*, April 1980, p. 26.
[6] W. H. Sanford, Jr., "A Modest 45-Foot DX Vertical for 160, 80, 40 and 30 Meters," *QST*, September 1981, p. 27.

By Richard R. Schellenbach, W1JF   From *QST*, November 1982

The JF Array

You don't have a "green thumb" for antennas? This multiband "antenna farm" is easy to grow!

It is purely accidental that the name of this antenna and my call sign are identical. However, it is no accident that the JF Array is a relatively simple but highly effective antenna system. It covers the 80, 40 and 15-meter bands from a single transmission line. This antenna provides significant gain on the 40- and 15-meter bands, while acting as a standard $\lambda/2$ dipole on 80 meters. In fact, the array may be used on all hf amateur bands, without the gain and directional characteristics found on 40 and 15 meters.

The initials "JF" describe the physical configuration of this array. On 15 meters, the antenna consists of two back-to-back "J" type radiators; hence the name "J Flat-Top," which is shortened to JF.

Theory of Operation

In essence, the JF Array operates as four $1/2$-λ elements in phase on 15 meters, and two $1/2$-λ elements in phase on the 40-meter band. On both bands, the feed impedance is extremely high. Therefore, an open-wire feed line (300 to 600 ohms) is recommended between this antenna and your Transmatch. Remember that this is a balanced antenna system, so it is desirable to maintain current balance from the antenna all the way back to your matching network.

Under some circumstances you may find

> *The JF Array provides an inherent diversity effect on the higher frequencies because of the large capture area. This effect greatly reduces fading that may occur during certain propagation conditions.*

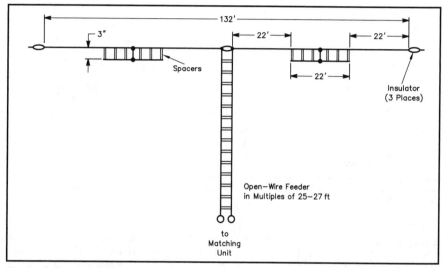

Fig 1—A dimensional drawing of the JF Array.

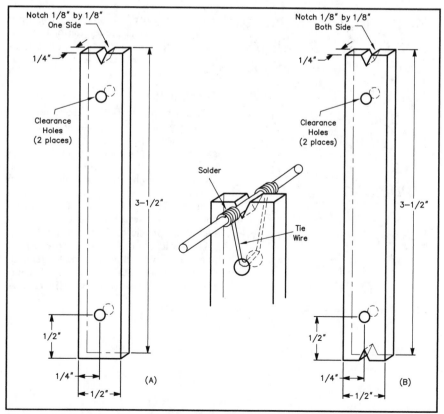

Fig 2—Spacer construction details for antenna sections are shown at A. Details of a spacer used for constructing the open-wire line are shown at B.

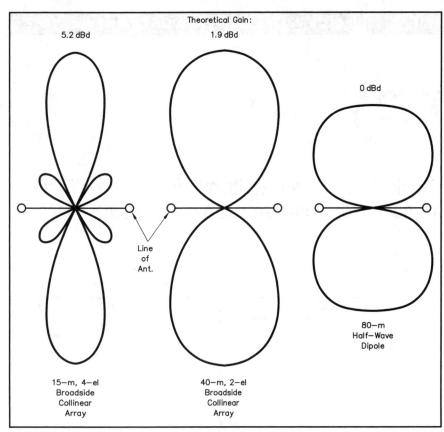

Fig 3—Radiation patterns and theoretical gain figures for the JF Array on different bands (for comparison only).

Under some circumstances you may find it necessary to experiment with the length of your open-wire feeder. This is because some operating frequencies and line-length combinations present a load impedance beyond the capability of your matching network.

it necessary to experiment with the length of your open-wire feeder. This is because some operating frequencies and line-length combinations present a load impedance beyond the capability of your matching network. The use of a nonharmonic-length feeder is the usual prevention or cure for this condition. Feeder lengths in multiples of 25 to 27 feet should allow all-band operation without any problems.[1]

Construction

The flat-top section of the antenna is made from no. 14 copperweld, or no. 12 hard-drawn copper wire (Fig. 1). Heavy-gauge wire is necessary to support the considerable weight of the array. The two stub sections should be made from no. 14 or 16 hard-drawn copper wire, and are held apart from the flat top by means of homemade spreaders. These are fabricated from 1/4-inch-thick plastic or Plexiglas sheet (Fig. 2A). The length of individual spreaders is not critical, but it should not be longer than 4 inches to prevent the wires from becoming unwieldly during installation. A

[1]m = ft × 0.3048.

The initials "JF" describe the physical configuration of this array. On 15 meters, the antenna consists of two back-to-back "J" type radiators; hence the name "J Flat-Top," which is shortened to JF.

spreader should be placed every foot, or less, along the stub to provide support and to prevent undue movement during windy periods. The spreaders are held to the main antenna wire by small lengths (4 inches) of no. 14 or 16 copper wire. This tie wire should be passed through the clearance hole at the "V" groove end and wrapped tightly on both sides (Fig. 2). The stub wires then are passed through the opposite clearance hole, and not tied, allowing freedom of movement for stress-free support. After attaching the stubs, solder a jumper wire between the center-tap of each stub and the main antenna wire (Fig. 1). Ensure good electrical connections by first scraping off any enamel insulation or oxidation, and by wrapping the wires tightly before soldering.

Balanced feeders may be purchased, or constructed from no. 14 or 16 copper wire spaced apart by the spreaders shown in Fig. 2B. Various types of commercial open-wire transmission line offer the builder a light-weight, already-built option. Any of the popular 300 to 600 ohm lines will do.

Performance

It is worth noting that the JF Array radiates the main power lobe broadside to the wire and not off the ends as a conventional, harmonically operated antenna does (Fig. 3). With a properly balanced feed line, you will observe that the array has an extremely clean radiation pattern. Installed at the 30 to 45-foot level, the antenna provides good DX performance. There is yet another desirable advantage to be found: the JF Array provides an inherent diversity effect on the higher frequencies because of the large capture area. This effect greatly reduces fading that may occur during certain propagation conditions. Give this simple antenna a try. You shall be pleasantly surprised!

Other Bands For The JF Array

I've had numerous inquiries concerning my article, "The JF Array" (Nov. 1982 *QST*). Readers wanted additional information about other bands of operation. Because of this response, I've shown applications of my array for other band combinations (Figs. 1, 2 and 3).

> *Higher gains are possible with the JF concept over conventional perpendicular phasing-stub arrangements because the outer elements are wide-spaced.*

I also recommend that expansion of the JF or any other collinear array not continue beyond four elements. This is because current diminishes rapidly beyond four elements, and proper phase relationships become difficult to maintain. Also, note that higher gains are possible with the JF concept over conventional perpendicular phasing-stub arrangements because the outer elements are wide-spaced. The "clean" lines of the JF also provide improved multiband performance because little discontinuity is offered at frequencies other than what the antenna was designed for. —*Dick Schellenbach, W1JF*

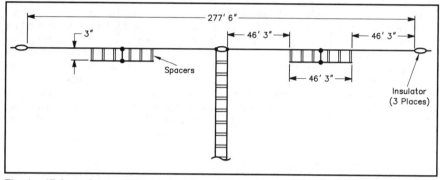

Fig. 1—JF Array for the following bands and (theoretical gain) configurations: 30 meters— 4-element collinear (5.2 dBd); 80 meters—2-element collinear (1.9 dBd); 160 meters—$\frac{1}{2}$-wave dipole; 40, 20, 15 and 10 meters—extended dipole.

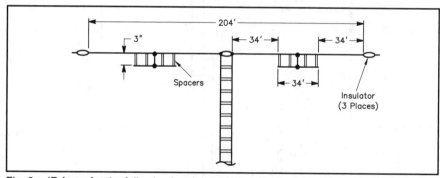

Fig. 2—JF Array for the following bands and (theoretical gain) configurations: 20 meters— 4-element collinear (5.2 dBd); 80, 40, 30, 15 and 10 meters—extended dipole.

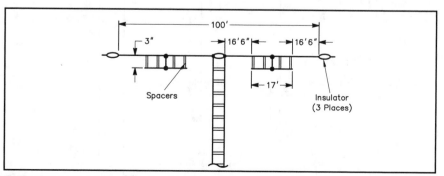

Fig. 3—JF Array for the following bands and (theoretical gain) configurations: 10 meters— 4-element collinear (5.2 dBd); 30 meters—two $\frac{1}{2}$-wave elements in phase (1.9 dBd); 40, 20, 15 and 10 meters—extended dipole.

From *QST*, September 1983 (Technical Correspondence)

The Quad-J-Collinear Antenna

This new antenna is essentially a four-element, wide-spaced, collinear broadside array. However, it has been reduced to the minimum possible overall size without diminishing its performance. The Quad-J-Collinear antenna is a highly effilcient, bidirectional array that produces a moderate 6-dB gain over a $1/2$-λ dipole. lf it's installed so the lower element is at least $1/2$ λ above ground, this array will provide low angle, elliptically polarized radiation useful for DXing. The elliptical-polarization feature provides a desirable polarization-diversity effect, which overcomes signal fading caused by constant variations in ionospheric refraction.

Fig. 1 illustrates the basic principle of operation, and may also be used as a design guide. Fig. 2 provides the dimensions for a 10-m antenna. The array is constructed from copper wire for light weight and low wind resistance. It may be suspended by nylon ropes from existing towers, masts or even trees—provided that sufficient separation between supports exists.

This is a fairly broadband type of array and provided the dimensional proportions are maintained, it will work effectively over a large frequency range. Overall bandwidth is primarily governed by the "J" type phasing stubs and the method employed in transmission-line matching.

Feed methods are left to your particular station requirements. However, it should be known that the feed point has a high impedance (>1 kΩ), and is balanced. Because of these characteristics, either tuned feeders or a $1/4$-λ matching section with a balun (Fig. 2) may be used. The length of coaxial cable from the balun to the transceiver is non critical in this application.—*Richard Schellenbach, W1JF*

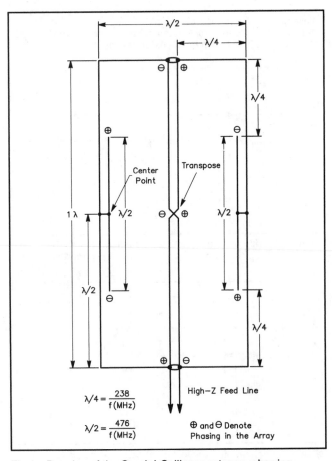

Fig 1—Drawing of the Quad-J-Collinear antenna, showing principal phase and element-length relationships.

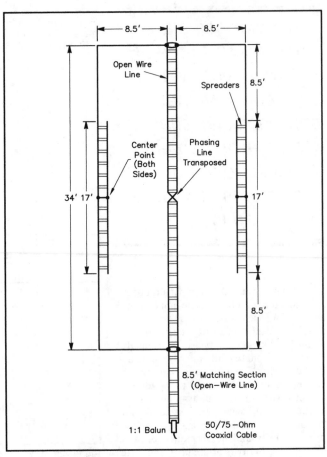

Fig 2—Dimensional drawing of a Quad-J-Collinear antenna for the 10-m band. Resonance is set for 28.0 MHz.

The J² Antenna For 10 and 24 MHz

This J² antenna was developed to cover the 10-MHz band, but with an eye toward future operation on the 24.89 to 24.99-MHz band. The antenna provides omnidirectional low-angle radiation with a single feed point. Fig. 1 shows the antenna dimensions.

On 10 MHz, the J² is configured as a $5/8$-λ vertical, which exhibits a theoretical gain of about 3 dB compared to a $1/4$-λ vertical. At 24 MHz, the J² becomes two in-phase, half-wave J antennas. The antenna base should be mounted not more than 2 feet above ground for best performance. You should provide a few $1/4$-λ radials for 10-MHz operation (23 feet). No radials will be required for the 24-MHz band, when that one becomes available for amateur use.

Matching to the base of the J² can be implemented with either an open-wire

> *Matching to the base of the J² can be implemented with either an open-wire transmission line and matching network in the shack, or by means of an L network at the base of the antenna.*

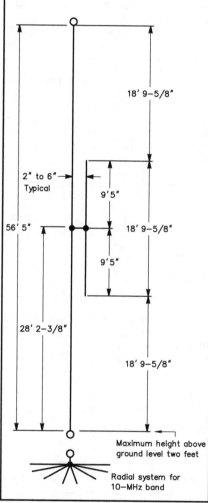

Fig 1—Dimensions and construction information for the W1JF J² antenna.

18' 9-5/8"

2" to 6" Typical

9'5"

56' 5"

18' 9-5/8"

9'5"

28' 2-3/8"

18' 9-5/8"

Maximum height above ground level two feet

Radial system for 10—MHz band

transmission line and matching network in the shack, or by means of an L network at the base of the antenna. The feed-point impedance will be high on both bands (>1000 ohms).

The antenna can be suspended from the side of a tower or from the limbs of a tall tree. Remember that both ends of the antenna are high-impedance points, so the rf voltage will be high. Use good insulators to support the main vertical wire.

The $1/4$-λ stubs are held away from the main wire by means of homemade Plexiglas spreaders. The length is not critical up to a maximum of about 6 inches. Position the spacers about 1 foot apart along the stubs, to maintain an even spacing.

This antenna is a little short of being $1/2$-wavelength long on 40 meters. By switching in some additional inductance at the base of the antenna, you should be able to use the J² on that band also. Operation as a $1/4$-λ vertical for 80 meters should also be possible, but that would require a much more extensive radial system.—*Richard Schellenbach, W1JF*

A High-Gain Monoband Directional Antenna

The X-ray antenna system was developed for the purpose of obtaining a simplified, high-gain antenna system with directional characteristics that can be changed quickly by remote control. This system provides:

• 6.5-dBd main-lobe gain
• broad bandwidth
• simple coaxial-cable feed
• instantaneous remote beam control
• low-angle DX capability
• modest height and space requirements

Essentially, the X-ray antenna consists of a pair of back-to-back, parallel connected, 1.25-λ V arrays. It could be described as an "inside-out" rhombic. This arrangement requires five supports, including a central support that is at least 0.5 λ high. The remaining four supports, however, can be significantly shorter because a 22° to 35° tilt is applied to all four antenna elements. The basic 1.25-λ V antenna presents an impedance of approximately 100 Ω at the feed point. Therefore, two of them connected in parallel result in an impedance that is near 50 Ω. This offers an extremely convenient point to apply coaxial-cable feed to the system. A 1:1 balun transformer preserves antenna balance. Also, a relatively broadband effect results from the wide-band balance and combined terminal impedances of both Vs.

The tilt angle, α, applied to each element provides a lower vertical-lobe angle. This favors long-range communication paths although it also produces quasi-elliptical polarization. Angle α, as shown in Fig. 1, may be anywhere between 68° and 55° with respect to the central support. Fig. 1 shows a plan and elevation view of the X-ray system.

Fig. 2 illustrates the method employed for relay control of directional characteristics, and Fig. 3 shows the relay enclosure mounted on the central mast.

The relay box contains a DPDT relay and a 1:1 balun transformer. Appropriate chassis connectors are mounted at the bottom of the box for remote relay control and coaxial cable to the station. The box should be waterproofed by covering any openings or seams with RTV (General Electric or Dow) sealant. Four ⅝- to 1-inch-diameter holes are provided at four corners of the enclosure and covered by square pieces of Lucite or Plexiglas sheet bolted to the box. Flexible wire leads, from the relay contacts, are brought out through small holes in the insulating sheets and connect to the antenna elements, as shown in Fig. 3.

Table 1 provides the antenna element lengths and the minimum recommended central-mast height for each operating band. The element-length formula is:

$$L = \frac{1230}{f} \qquad \text{(Eq. 1)}$$

where
L = length in feet
f = frequency in megahertz

The system completely covers all but the 10-m band, where it should be limited to any one 0.5-MHz segment. Cut the antenna elements for the center of any band shown in Table 1. Performance should be satisfactory, as long as all elements, including the leads to the relay and balun, are the same length.
— *Richard R. Schellenbach, W1JF, Reading, Massachusetts*

[1]mm = in × 25.4; m = ft × 0.3048

Table 1
Element and Mast Dimensions

Band	Element Length	Center Mast
40 m	172 ft	67 ft
30 m	122 ft	46 ft
20 m	87 ft	34 ft
15 m	58 ft	22 ft
10 m (28.0-28.5 MHz)	43 ft 6 in	17 ft
10 m (28.5-29.0 MHz)	42 ft 9 in	
10 m (29.0-29.5 MHz)	42 ft	

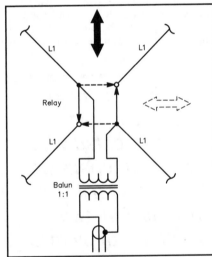

Fig 2—Schematic diagram of antenna array and switching system. The arrows indicate main-lobe orientation.

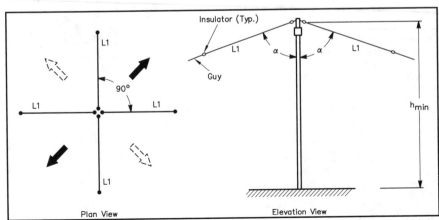

Fig 1—Plan View and Elevation of the X-ray antenna array. Minimum center mast height, h, is given in Table 1.

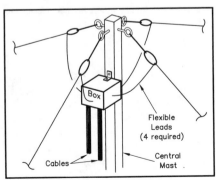

Fig 3—Arrangement of antenna and switching box at the top of the center mast.

NOTES

NOTES

NOTES

NOTES

FEEDBACK

Please use this form to give us your comments on this book and what you'd like to see in future editions, or e-mail us at **pubsfdbk@arrl.org** (publications feedback). If you use e-mail, please include your name, call, e-mail address and the book title, edition and printing in the body of your message. Also indicate whether or not you are an ARRL member.

Where did you purchase this book?
  ☐ From ARRL directly   ☐ From an ARRL dealer

Is there a dealer who carries ARRL publications within:
  ☐ 5 miles   ☐ 15 miles   ☐ 30 miles of your location?   ☐ Not sure.

License class:
  ☐ Novice   ☐ Technician   ☐ Technician Plus   ☐ General   ☐ Advanced   ☐ Amateur Extra

Name _____   ARRL member? ☐ Yes ☐ No

_____ Call Sign _____

Daytime Phone ( ) _____ Age _____

Address _____

City, State/Province, ZIP/Postal Code _____

If licensed, how long? _____ e-mail address _____

Other hobbies _____

_____

Occupation _____

For ARRL use only	WAC
Edition	1 2 3 4 5 6 7 8 9 10 11 12
Printing	9 10 11 12

From _____

_____

_____

EDITOR, ARRL'S WIRE ANTENNA CLASSICS
AMERICAN RADIO RELAY LEAGUE
225 MAIN STREET
NEWINGTON CT 06111-1494

— — — — — — — — — — — — — — — — — please fold and tape — — — — — — — — — — — — — — — —